Sexual Subjects

Sexual Subjects
Young people, sexuality and education

Louisa Allen

First published 2005 by
PALGRAVE MACMILLAN
Houndmills, Basingstoke, Hampshire RG21 6XS and
175 Fifth Avenue, New York, N. Y. 10010
Companies and representatives throughout the world

PALGRAVE MACMILLAN is the global academic imprint of the Palgrave Macmillan division of St. Martin's Press, LLC and of Palgrave Macmillan Ltd. Macmillan® is a registered trademark in the United States, United Kingdom and other countries. Palgrave is a registered trademark in the European Union and other countries.

ISBN-13: 978–1–4039–1283–1
ISBN-10: 1–4039–1283–1

This book is printed on paper suitable for recycling and made from fully managed and sustained forest sources.

A catalogue record for this book is available from the British Library.

Library of Congress Cataloging-in-Publication Data
Allen, Louisa
    Sexual subjects : young people, sexuality, and education / Louisa Allen.
        p. cm.
    Includes bibliographical references and index.
    ISBN 1–4039–1283–1 (cloth)
        1. Sex instruction for youth. 2. Youth–Sexual behaviour.
3. Sex instruction for youth–New Zealand. I. Title.

HQ35.A615 2005
613.9′071′293–dc22   2005042903

10   9   8   7   6   5   4   3   2   1
14   13   12   11   10   09   08   07   06   05

Printed and bound in Great Britain by
Antony Rowe Ltd, Chippenham and Eastbourne

For Andrew, with love

Contents

List of Tables

Acknowledgements

The research upon which this book is based was made possible by the financial support of the Health Research Council of New Zealand to whom I am eternally grateful. Thanks also to the Foundation for Research Science and Technology for granting me a post-doctoral fellowship and subsequently the literal and temporal space to write this book.

My development as a researcher has been expertly supported and cultivated by a handful of cherished mentors. Thank you Madeleine Arnot and Sue Middleton for lighting the way during my doctoral studies with astute analyses and infinite energy. Kay Morris Matthews and Roger Dale for sagacious advice and unfailing support always, and Alison Jones for inspiring a passion for feminist theories of education and providing me with more opportunities than I can name.

Through the tunnels and turns of this research I have encountered people whose dedication to sexuality education and professionalism has greatly impressed me, Sally Hughes you are one of these people whose wisdom has been invaluable.

Although I would like to name them individually ethical regulations dictate I cannot. Collectively then, I wish to thank all those energetic teachers and community training facilitators whose interest in my work enabled me to recruit a sample and engendered many a happy day in the field.

This research could not have been undertaken without the participation of the young people with whom I came into contact. For those who took part in the individual and couple interviews I would like to extend a special thank you for your candid responses, your time and your humour.

Thank you also to Connie Chai who helped prepare the draft manuscript and whose attention to details and type-setting skills I really appreciate.

It always seems that in acknowledgements those who have provided the most prolonged and often complex sustenance to a project are left to the last. I would like to thank my two treasured parents who have been instrumental in my lifelong education; My mother for instilling in me a belief in the power of women, the importance of social justice and teaching me that sexuality could/should be beautiful; My father for always being a pillar of strength, with an unfaltering belief in anything I do and ability to provide exactly the right sort of emotional/ practical support and injection of humour that rights my world.

And to Andrew who knows me best, and brings light and joy into my everyday life. Thank you for making me laugh, and for the warmth of your love that lifts me out of daily struggles. It is these things that make my life a pleasure.

1
Introduction

Locating the research

I was recently asked what drew me to research sexuality. My answer was that this area carries a social stigma that renders talk about sexuality and its pleasures often uncomfortable and sometimes perverse. This constitution of sexuality conflicts with my view that it is a potential source of positive energy and pleasure, about which there is nothing inherently embarrassing or dirty. Instead, pleasure adds meaning to our lives (Tepper, 2000) and is a defining feature and motivating factor of social action. Feminists have often argued that the ultimate purpose of research is 'to change the world, not only to study it' (Stanley, 1990, p. 14). The research discussed in this book has evolved out of my own sense of a need for change with regard to the social constitution of sexuality and its institutional capture in sex education.[1]

A main concern of this work has been to gain greater insight into young people's (hetero)sexual subjectivities, knowledge and practices and to think about how such understandings might inform sexuality education. This task has involved understanding young people's own conceptualisation of their (hetero)sexual selves, knowledge and practices and what these imply for how we conceptualise sexuality education's effectiveness. My exploration of these themes is intersected by an interest in gender and power and the way this is implicated in our constitution of sexual subjectivities, knowledges and practices. The study is framed by a feminist methodology and employs the tools of post-structuralism to make meaning from these data. Fuelled by a desire 'to change the world' this book aims to reshape understanding around young people's (hetero)sexualities and propose new ways of thinking about how we teach about sexuality.

1

Personal motivations are inextricably linked to wider social processes and structures. While this study emerges from my concern to critically reflect upon youth and (hetero)sexuality, it is also structured by a particular international and local social context. One global variable here is the spread of HIV/AIDS which first incited moral panic in countries like the UK, US (Weeks, 1989) and New Zealand (Davis, 1996) in the 1980s. As Holland, Ramazanoglu, Sharpe and Thomson (1998) explain, 'once HIV/AIDS was identified as a fatal condition that could be transmitted through sexual activity, it aroused potent fears of death which expressed confusion and uncertainty about sexual behaviour and identities' (p. 1). Amid this wave of anxiety it became apparent that any bid to curb this disease must entail greater insight into people's sexual knowledge and practices. While the study of sexuality in Western countries has a rich history dating back to the European sexological pioneers of the late nineteenth century, the HIV epidemic generated renewed interest and opportunities for this work (see Weeks, Holland, and Waites, 2003 for fuller discussion). Attention turned to youth with the realisation that HIV may be contracted during early sexual encounters (Aggleton, Ball, and Mane, 2000). Investigating how young people's sexual behaviour could put them at risk of HIV formed part of a volume of research concerned with 'the social aspects of AIDS' (Aggleton, Homans, and Warwick, 1988).[2] The current study owes much to issues and theories which emerged out of this research context especially those concerning gender, power and the negotiation of safer sex (Holland et al., 1998; Waldby, Kippax, and Crawford, 1993; Weeks and Holland, 1996; Aggleton, Homans, and Warwick, 1988; Redman, 1996).

While HIV/AIDS forms part of the global climate in which this project was formulated, local conditions have also shaped its conceptualisation. Concern over increasing rates of other sexually transmissible infections (STIs) and what is constituted as the 'problem of teenage pregnancy' have also been influential. In 1996 the year I applied for funding for this study, the then Minister of Health Jenny Shipley was courting media and public attention around sexuality issues in the form of a Sexual and Reproductive Health Strategy. Centred firmly on combating sexually transmissible infections and teenage pregnancy this strategy aimed 'to promote responsible sexual behaviour, to minimise unintended pregnancies, reduce abortion rates, and the incidence of sexually transmitted diseases and HIV/AIDS' (Ministry of Health, 1997, p. 1). STI's are problematic for governments because they can 'lead to infertility from pelvic inflammatory disease, cancer and other chronic diseases' all of which take an economic and social toll on the state (Ministry of Health, 2001, p. 1). Encapsulated within the introductory statement of a more recent sexual and reproductive health strategy, teenage pregnancy is conceptualised as a problem because it '... not only creates more social and economic problems for the mother, but also increases the child's risk of poor outcomes in education, health and welfare' (Ministry of Health, 2001, p. 1). New Zealand is not the only country to construct this view of teenage pregnancy, with similar concerns being expressed by The New Labour Government in Britain. New Labour has argued that academic achievements and effective labour market participation are inhibited by early or 'premature' parenthood (Social Exclusion Unit, 1999). However, the extent to which teenage pregnancy is seen as a problem is contingent upon its social constitution as such (see Alldred, David, and Smith, 2003, for a fuller exploration of this idea). The naming within social policy of teenage pregnancy and STIs as a crisis, has created opportunities and legitimated projects like the current research to investigate young people's sexual subjectivities, knowledge and behaviour.

A climate of concern around these consequences of sexual activity has shaped this study as an exploration of what is conceptualised within the sexuality education literature as a knowledge/practice gap. This describes the way in which what young people learn in sexuality education, about for example safer sex or how to prevent pregnancy is not always put into practice. The notion of a gap has emerged from research which has attempted to evaluate the effectiveness of sex education's messages by equating this with a reduction in STIs and unplanned pregnancies. Studies documenting examples of the gap phenomenon have revealed that despite recognising that condoms can prevent STIs and unplanned pregnancies and knowing how to employ them, some young people still use them inconsistently (Molitor, Facer and Ruiz, 1999; Hingson and Strunin, 1992; SSRU, 1994; Ogden, 1996; Wellings and Field, 1996). When I began this research as part of my doctoral studies, my aim was to think critically about the gap phenomenon and why it might occur. I decided to examine the variables I felt were implicated in its existence that is, young people's sexual knowledge, their sexual practices and their sexual subjectivities, and gain a sense of how young people themselves conceptualised these things and possible relationships between them. As a result of structuring the project in this way it became apparent that young people's own understandings of why knowledge gained from sexuality education was not always translated into practice, was not conceptualised as a gap. The gap appeared to be a creation of academic researchers and sexual health professionals which did not capture the nuances and complexities with which participants conceptualised their sexual knowledge, practices and selves. In order to give recognition to young people's own understandings of this phenomenon, the word 'gap' appears in quotation marks throughout this book. A central purpose of this research was to elucidate young people's sense of what their sexual knowledge, practices and subjectivities are and in this way problematise the notion of the 'gap' phenomenon for determining sexuality education's effectiveness.

Sexuality education: The New Zealand context

Historically sex education has served as a vehicle for furthering a number of moral and social imperatives, of which arming young people with knowledge about how to prevent STIs and pregnancy has endured (Willig, 1999). While a comprehensive survey of sex education is available elsewhere (Smyth, 2000), here I want to highlight some of the historical features which have shaped this school subject and its philosophy. Smyth's account of contraception, sex and politics in New Zealand reveals that the medium of early sex education was pamphlets and manuals influenced by religious and Eugenicist ideas. These publications expounded the dangers of masturbation and the need to sustain a 'fit' New Zealand population (Smyth, 2000). Eugenicists viewed sex education or sex hygiene as it was also known, as a strategy to further their aims of securing a 'pure race'. For faith based organisations sex education was a means of preaching about a proper relationship with God where expressions of sexuality were confined to marriage. During the First World War the government recognised the positive role sex education might play in curtailing venereal disease, when there was an escalating incidence amongst soldiers. The success of Ettie Rout's campaign to curb venereal disease with education and condoms paved the way for what has become one of sex education's underlying aims, disease prevention.

From its inception into public debate in 1912 sex education has always been surrounded by controversy. Early discussions were preoccupied with its presence in schools, a debate which appeared to ebb and flow with the submission of a series of reports on this topic.[3] Due to the way sexuality issues have been socially constituted as private, embarrassing, unspoken and wrong, recommendations to allow sex education in schools and make it compulsory have often met with virulent opposition from the moral right. Ryan (1988) describes the 'moral right' as encompassing a number of different interest groups comprising of religious organisations and churches. Although they have diverse organisational bases, membership strategies and objectives, one of their defining features is their focus on the 'family' who they perceive should rightly teach children about sex. While the moral right has always constituted a proportionately small sector of the general population they have been vociferous, winning media attention with cleverly crafted sound bytes which are emotive and sensationalised. Their campaigns have made sex education politically sensitive for governments to openly support, contributing to a situation in which sexuality education is the only curriculum subject to have an out-clause. This means that parents and caregivers can specify that their children be withdrawn from classes with content about sex and sexuality.

As it currently stands, sexuality education is a component of health education (Tasker, 2004). In 1999 a new *Health and Physical Education* curriculum was released replacing the previous health, physical education and home economics syllabuses. Sexuality Education was named as one of the key areas of learning in this curriculum with attention drawn to a distinction between this and the concept of 'sex education'. According to the curriculum:

> the term 'sexuality education' includes relevant aspects of the concept of hauora,[4] the process of health promotion and the socio-ecological perspective.[5] The term 'sex education' generally refers only to the physical dimension of sexuality education' (Ministry of Education, 1999, p. 38).

The curriculum states that the overall purpose of sexuality education is to 'provide students with the knowledge, understanding, and skills to develop positive attitudes towards sexuality, to take care of their sexual health and to enhance their interpersonal relationships, now and in the future' (Ministry of Education, 1999, p. 38). While the *Health and Physical Education* curriculum is compulsory up to the end of year 10 (second year of secondary school) special legislative provisions regulate the sexuality education component. These stipulate that under section 105C of the Education Act 1964 the principal is required to make a written report to their board of trustees[6] following consultation with the school's community. Subsequently, the Board of Trustees can 'direct, or refrain from directing' the school to include in the teaching programme 'any particular element of sex education described in that written description'. This clause gives potential power to the school community to determine what might be included in a sexuality education programme. As mentioned above section 105D 'provides for parents and caregivers to apply to the principal in writing, to have their child to be 'excluded from every class in which any element...that is sex education is being taught' (Ministry of Education, 1999, p. 39).

One of the consequences of these provisions is that it is the school's discretionary capability that dictates whether young people receive sexuality education after year 10 (students who are approximately 14 years). This means that during the period of average age of first sexual intercourse in New Zealand (17 years) many young people will not receive any sexuality education unless they choose health as a subject (Dickson, Paul, Herbison, and Silva, 1998). Despite efforts to raise the academic status of this curriculum there is a common perception amongst students with intellectual aspirations that this subject is not academic enough. A large proportion of the senior school population which might benefit from sexuality education's content choose options they believe will better pave their way to a university and/or professional career. There is also an issue in these legislative regulations around the rights of young people to access information they deem important. For instance, what if a student wanted to attend sexuality education classes because they felt the sexuality education they were receiving at home was inadequate, but their caregiver had made an application for them not to attend? Under the current regulations there is a possibility that young people could be denied the knowledge and participation they desire and need. Such tensions are especially apparent for young people who prescribe to contemporary New Zealand youth culture but whose families have deeply held traditional cultural and religious beliefs (for example, refugees from middle eastern countries, first generation New Zealanders from Pacific nations, Asian students who have come to study from Taiwan, Korea, China, etc.).

Other ministerial policies and initiatives have shaped the recent content and delivery of sexuality education in New Zealand. In 1996 as part of the Sexual and Reproductive Health Strategy mentioned above, the Education Review Office (ERO) conducted a review of the sexual and reproductive health components of the curriculum. As a result of this report, a new National Curriculum Statement of Health and Physical Education was created (Ministry of Education, 1999) and the issues it raised have since occupied the government's sexual health agenda. Of significance were the findings that few sexuality education programmes at secondary school were as long as the 14 hours per year recommended by research as essential for effective education in this area, and that most school staff delivering such programmes were inadequately trained. In addition, few schools reviewed their reproductive and sexual health education programmes in order to increase their efficacy and only half undertook the required consultation process with parents and caregivers.

The new *Health and Physical Education* curriculum has endeavoured to address sexuality education's pedagogy while the *Sexual and Reproductive Health Strategy* has provided a vision of young people's sexual and reproductive health. Main features of the strategy have been to encourage young people to delay the onset of sexual activity and improve access to contraceptive information and products through the reduction of cost barriers (Ministry of Health, 1997). Within this conceptualisation of young people's sexual health, what is prioritised is their being 'disease and child free'. This sentiment is echoed in the 2001 sexual and reproductive health strategy where the government's concerns are again outlined as 'sexually transmitted infections and the high level of unintended/unwanted pregnancies' (Ministry of Health, 2001, p. 1). Despite rhetoric which acknowledges 'positive sexual identity and sexuality as fundamental to our sense of self, self-esteem and ability to lead a fulfilling life', what appears to define and dominate this strategy is reducing STI's and unintended/unwanted pregnancies. The naming and framing of the strategy in terms of 'sexual and reproductive health' reflects this continued emphasis, continuing to pair sexual expression with reproduction. It is within this environment that current sexuality education programmes are implemented and this book's thinking about young people as sexual beings is conceptualised.

Conceptual frameworks

This research engages with a number of key issues and concepts within feminist, sociological and post-structural thinking. While these are explored in greater detail throughout the ensuing chapters, here I want to highlight the main concepts and frameworks which constitute the theoretical underpinnings of the project. Sexuality is a central concept in this book and deployed in accordance with post-structural understandings of this term. Like Epstein, O'Flynn, and Telford (2003) sexuality is understood as,

> ...something much more broadly understood than simply 'sex' or 'sexual relationships'. It is our premise that sexuality is not the

property of an individual and is not a hormonally or biologically given, inherent quality. Rather sexual cultures and sexual meanings are constructed through a range of discursive practices across social institutions including schools. Thus, when we talk about 'sexuality' we are talking about a whole assemblage of heterogenous practices, techniques, habits, dispositions, forms of training and so on that govern things like dating and codes of dress in particular situations (p. 3).

What is given precedence in this understanding of sexuality is the way it is actively constructed in particular contexts through various discursive practices. Sexuality is not simply a biological product of innate and immutable quality, but the consequence of social practices which are infused by power and mutable. This discursive constitution means that although sexuality is experienced by subjects as personal and emanating from within, it is not individually produced. On the other hand socially constituted subjects are not devoid of agency in the way sexuality is 'lived'. The extent and nature of young people's sexual agency (discussed below) occupies significant space in this book. A further consequence of conceiving sexuality as discursively constituted and intersected by relations of power is that sexualities can be seen to be hierarchically organised (Weeks, Holland, and Waites, 2003). Sexualities are produced through social structures like gender, class, age, ethnicity and physical ability in ways that render some forms dominant and others subordinate. An overarching consequence of this structuring of sexualities is the institutionalisation of heterosexuality and marginalisation of other configurations of sexual identity. The need to redress this sexual inequality is a motivating principal behind the argument progressed by this research for a discourse of erotics in sexuality education.[7]

Another conceptual understanding employed in this research emerges from sociological work on youth (Willis, 1977; Mac an Ghaill, 1996a; McRobbie, 1996; Kehily, 2002) and a recognition that young people are active (sexual) agents. This view has developed in response to the perception that young people's relatively short life experience compared to adults means they are less capable of making the 'right' decisions. This sense of young people's lack of potency is mirrored in the idea that school students are empty vessels to be filled with knowledge and wisdom delivered by the teacher. These sorts of conceptualisations negate the fact that youth already possess particular knowledges and have their own aspirations, agendas and group commitments (Johnson cited in Kehily, 2002, p. xiv). Drawing on the work of Aggleton Aggleton et al., the current research attempts to incorporate the kind of sentiment communicated in the following extract.

> What is so badly needed is a new vision of young people as harbouring the potential to shape their own lives, rather than as troubled, or as trouble makers. Only by providing young people with the respect and understanding they deserve, and by listening carefully and non-judgementally to what they have to say, can we as helping professionals play our proper part in creating a healthier and more age equitable future. (Aggleton, Ball, and Mane, 2000, p. 220)

This research was structured in terms of not just an exploration of young people's sexual knowledge and practice but also how they perceived themselves as sexual subjects. The latter aim acknowledges that any notion of a knowledge/practice 'gap' must involve the agency of subjects who action knowledge into practice. My desire to constitute young people as active and productive social agents also prompted this examination of the 'gap' phenomenon from young people's *own* perspectives. In asking how they understood their sexual knowledge, practices and subjectivities and the potential relationships between these, I hoped to prioritise their own conceptualisations. The centring of young people's needs and interests is also incorporated in my hopes for the way these might inform design and policy around sexuality education programmes. As Kehily (2002) notes about her work on student cultures, this focus can be seen as a way of '"giving voice" to school students who receive the curriculum but play no part in the structuring of the school as an organisation or the planning of the curriculum and the teaching of lessons' (p. 2). This 'voice' as expressed in the current research findings, is now informing the design of another research project to create a sexuality education resource for 16–19 year olds.

Integral to any notion of agency is a theory of the subject. In conceptualising this research as including an analysis of young people's (hetero)sexual subjectivities I draw upon the theoretical insights of post-structuralism. According to Henriques, Hollway, Urwin, Venn, and Walkerdine (1984), 'subjectivity' refers to 'individuality and self awareness – the condition of being a subject' (p. 106). My interest in subjectivity focuses on young people's understandings of themselves as sexual subjects in relation to others and the world (Weedon, 1987). For feminist post-structuralists, forms of subjectivity are the product of discourses or ways of constituting meaning which are historically located and hence mutable. The intersection of power with discourse means that the individual is always the site of conflicting forms of subjectivity, which results from the fact that:

> As we acquire language, we learn to give voice – meaning – to our experience and to understand it according to particular ways of thinking, particular discourses, which pre-date our entry into language. These ways of thinking constitute our consciousness, and the positions with which we identify structure our sense of ourselves, our subjectivity. (Weedon, 1987, p. 33)

Particular discourses are imbued with more power to constitute the subject through their entrenched institutional locations, and therefore offer a preferred way of understanding ourselves. For example, one understanding of femininity is that because women have a biological propensity to give birth they are naturally more nurturing than men. This idea is contained and reproduced within discursive fields such as the law (where mothers are given presiding authority over their children) and the labour market, where provisions for maternity leave and lack of gender pay parity favour women taking primary responsibility for children's daily care. The fact that there are also competing discourses about femininity means that women might understand their role in ways that sit in tension with these understandings (for example, as the sole wage earner). It is in the contradictory nature of subjectivity that spaces for agency occur enabling individuals to take up subject positions which may seem to address their interests more directly (Weedon, 1987). The chapter on young people's sexual subjectivities explores this idea further and reveals agency is not a simple case of 'choice' about how we understand ourselves as sexual people. This sense of self is always limited by language and by the particular discourses to which we have access.

Another concept which provides a theoretical premise from which discussion is launched is (hetero)sexuality. Feminist and queer theories have sought to problematise this term and therefore I employ it with a recognition of what Ingraham (2002) has called 'the heterosexual imaginary'. This concept describes:

> the way of thinking which conceals the operation of heterosexuality in structuring gender and closes off any critical analysis of heterosexuality as an organizing institution. The effect of this depiction of reality is that heterosexuality circulates as taken for granted,

naturally occurring, and unquestioned, while gender is understood as socially constructed and central to the organisation of everyday life. (p. 79)

What Ingraham suggests is that (hetero)sexuality is so deeply embedded within accounts of social and political participation, that it is seen as the normative and natural category against which the sexual 'other' is defined. This naturalisation of (hetero)sexuality means that within social relations it is taken for granted rather than overtly acknowledged or problematised. It is in recognition that (hetero)sexuality is not synonymous with sexuality nor is there anything natural about its expression that 'hetero' appears in brackets throughout. Instead, (hetero)sexuality is referred to as a structuring institution and set of practices which organises the regulation of relations between men and women. It is intimately tied to gender in the way that it depends on gender divisions for its meaning (Richardson, 1996). For example, (hetero)sexual desire is constituted through a gender arrangement presuming that if you are a man you will inevitably be attracted to a woman and vice versa. This attraction occurs because conceptualisations of gender contain an assumption of heteronormativity where appropriate masculinity/femininity is measured by their relation to a constituted opposite. To be 'truly' masculine then, is to display desire for and sexual interest in women. Evidence of this gender/sexual order is revealed in Chapter 6, where participants explain how having a relationship made them feel more masculine or feminine.

This book is also in dialogue with feminist debates around the operation of power within (hetero)sexual relationships. Some feminists have characterised the power which mediates (hetero)sexual relations dualistically as male dominance and female submission (Jeffreys, 1996; Mackinnon, 1989). Within this framework (hetero)sexual desire is organised around eroticised power difference where 'difference' between genders is thought to elicit sexual excitement (Jeffreys, 1996). One of the problems with this conceptualisation is that it portrays male power as monolithic, conceding minimal agency to women. This paradigm proposes a repressive type of power which does not explain why women and men might engage in and enjoy these relationships, or how some women perceive themselves as exercising power within them. These complexities around power and agency are contained within Foucault's (1976) question 'Would power be accepted if it were entirely cynical?' (p. 86). He suggests here that 'power is tolerable only on condition that it masks a substantial part of itself', so that subjects 'see it as a mere limit placed on their desire, leaving a measure of free-dom, however slight- intact' (p. 86). Throughout this book I explore Foucault's ideas about the productive capabilities of power to produce thoughts and action and a sense of agency (however slight). From a Foucauldian perspective, I take power to mean a shifting nexus of rela-tions that act web-like through institutions, practices and material sub-jects 'without being exactly localized within them' (Foucault, 1976, p. 96) and where the possibility of resistance is always present. In the context of young people's relationships this means endeavouring to understand how young women and men are constituted as sexual agents and where agency is exercised (especially by young women) in (hetero)sexual encounters.

Structure of the book

In its exploration of the 'gap' phenomenon this book reveals a struc-ture consistent with its central themes around young people's sexual subjectivities, knowledge and practices. Before embarking on an explo-ration of these it commences with an examination of some of the methodological complexities of conducting research on youth and sex-uality in schools. The first part of the chapter provides an explanation of the methodological framework, methods and analysis employed in order to define the projects' parameters and limits. This discussion is followed by a more indepth analysis of methodological decisions around the issue of participants' ethnicity which impacted upon the research's production of knowledge. As a means of establishing the materiality of its own production, weaved throughout this chapter is an examination of how my subjectivity has shaped the process and products of this project.

Chapter 3 undertakes an examination of participant's conceptualisa-tions of their sexual knowledge contrasting these with how sexual knowledge has traditionally been constituted within sexuality educa-tion programmes. The purpose of this chapter is not only to reveal how participants understood their sexual knowledge, its possible relation-ship to sexual practice and the sources from which it was derived, but also to problematise conventional perceptions of this within sexuality education curricula. What I suggest is that the kinds of knowledge traditionally offered about sexuality in schools is not the sort of knowl-edge these young people were most interested in, or held in highest esteem, and that this has implications for sexuality education's design and delivery.

The title of the next chapter, 'Sexual Subjects' reflects its underlying argument about the need to recognise young people as *being sexual* and *exercising agency*. This agency is seen in the way that young people draw not only on dominant discourses of (hetero)sexuality in their talk about themselves as sexual people but also resistant discourses. What is revealed here is how participants' understandings of sexual subjectivity are gendered and have particular ramifications for the possibilities of sexual practice. The chapter also explores young people's sexual agency in terms of how they actively manage their sexual identities within the research context. Through their talk about sexuality, participants engage in what I explain as 'identity work' evidenced in the things they choose to reveal and conceal about their sexual selves. In the final section of this chapter I consider how these insights about the gendered nature of young people's sexual subjectivities and their constitution as active sexual agents are useful for thinking through the 'gap' phenomenon.

As a means of extending our conceptualisation of sexual subjectivity to incorporate corporeality, Chapter 5 analyses young people's sexual embodiment. Considering the pleasurable and desiring body in relation to the 'gap' phenomenon draws attention to the body's centrality in sexual encounters. I set out to describe how young people speak about the experience of their bodies during sexual activity and in so doing endeavour to extend general theories of embodiment to the realm of the sexual. In the context of thinking about the 'gap' phenomenon, my concern is to advocate for, and create a political theory of, sexual embodiment. This recognises that sexual embodiment is not only gendered but that gendered bodies are subsequently given access to various configurations of power which result in particular positive and negative effects. This discussion *enfleshes* thinking around the knowledge/practice 'gap' and considers how gendered bodies might be implicated in this phenomenon.

As (hetero)sexual practice often involves the playing out of subjectivities within a relationship context Chapter 6 concentrates on this particular site. Given the paucity of empirical data about what young people's relationships look like (especially in New Zealand) the first section documents this in terms of their number length and type. For those who work with or care for young people, this kind of information is extremely valuable for contextualising and understanding their sexual experiences. The pleasures of young peoples' relationship experience are often considered obvious or unimportant and subsequently rarely explored in sexuality education content. This chapter documents them as conceptualised by young people themselves as a means of acknowledging their importance to participant's lives and contributing to the development of a discourse of erotics (see below). In order to analyse why knowledge may not always be translated into practice it is important to consider how power operates in relationships. The way power is sexualised in the relationship context and the effects this has for how participants' sexual subjectivities are 'lived' is examined. The significance of the final element of the chapter for understanding the 'gap' phenomenon is the way it discusses participants' own thoughts about how *knowledge* is or is not operationalised in *practice*.

Chapter 7 pulls the findings of the book together in an argument for the inclusion of what I have called a discourse of erotics in sexuality education. I do this by indicating how insights from each of the previous chapters point towards the need for such a discourse within sexuality education as a way of tackling the so called knowledge/practice 'gap'. A case for the possible benefits of making space for this discourse within sexuality education is made, as well as an explanation of how a discourse of erotics might take shape. As part of this discussion what a discourse of erotics might mean for particular segments of the youth population is considered. Such sectors include, gay, lesbian, bisexual, takataapui,[8] intersex and transgendered youth, young people of different cultural and religious backgrounds as well as young people with disabilities. While these young people were not targeted by the study, I argue their visibility is integral to understanding youth and sexuality. The inclusion of such issues of diversity in this book is a statement about how imperative any acknowledgement and inclusion of these young people are, to the future of sexuality education policy and programme design.

The final chapter draws together some of the main research findings and evaluates their implications for better ways of conceptualising the design and delivery of sexuality education. In addition some of the questions left unanswered by the research are highlighted and possible areas for future exploration identified. As the title of this chapter implies, this is not a definitive conclusion to the issues raised throughout but a springboard from which to launch new questions and make other discoveries. It is in this vein that I begin...

2
Researching Sexuality: Methodological Complexities

Locating the researcher

It is as a matter of politics that Chapter 1 commenced with a personal reason for researching sexuality. Beginning symbolically and literally with 'I' represents a challenge to traditional academic 'authority' where there is an absence of the author's personal voice in texts. Like other social scientists (Lather, 1991; Middleton, 1995; Jones, 1992; Hertz, 1997) I seek to locate myself within this project in recognition that research findings cannot be separated from their means of production and my own implication in this process. This narrative about young people and sexuality is not seamless, objective or in any way the whole picture of what it means to be a young and a sexual person in New Zealand. It is shaped by my own situated and partial perspective evidenced in the questions I chose to ask and the ones I missed out. Letherby (2003) explains that 'being reflexive and open about what we do and how we do it, and the relationship between this and what is known, is crucial for academic feminists as it allows others who read our work to understand the background to the claims we are making' (p. 6). This chapter establishes how the research was designed and the consequences of this for the knowledge produced. As the chapter unfolds, I reflect upon how my positionality has affected the formulation of research questions, relationships with people in the field as well as the analysis and interpretation of data. Before I can do this however, I need to introduce myself so you can see the place from which I stand/speak.

As this book is concerned with sexuality my construction of self begins with the fact that I am (hetero)sexual. Although this is the conventional terminology I use to describe my sexual identity, I prefer to

15

think of myself as what Thomas coins 'straight with a twist' (Thomas, 2000, p. 3). I am 'straight' in terms of being (hetero)sexual, however 'the twist' is my recognition of the fluidity and diversity of sexual identities and a political/theoretical interest and commitment to decentring (hetero)sexuality. Along with my (hetero)sexual identity I am also a 31 year old, Pakeha[1] woman who was born in England and emigrated to New Zealand at four years old. As a consequence of living most of my life in New Zealand and feeling a strong connection with its landscape and culture, I consider myself a New Zealander. I choose to describe myself as Pakeha rather than European, in acknowledgement of the historical presence of New Zealand's indigenous Maori population and the importance of Maori culture to this country's identity. I am also middle class, being the only daughter of parents with white collar occupations in the information technology and education sectors. My parents instilled a political consciousness in me from an early age, and as critical thinkers who read avidly, they questioned everything and believed categorically in social justice. Their influence and those of my primary and secondary school teachers who insisted that 'girls can do anything', meant that I have considered myself a feminist since the age of six. At this time I remember sitting on my teacher's knee and telling her resolutely that I did not want to be 'just a housewife', having somehow detected that this was a socially undervalued job to which mostly women were confined. Each of these aspects of identity find expression in this project's construction in ways that I am both aware and unable to 'see'. As explained in the sections below, my (hetero)sexuality meant I had limited access to same-sex attracted communities, my Pakeha ethnicity presented a challenge to working with Maori, while my feminist politics influenced the research topic choice and methodology. Throughout the rest of the chapter it is possible to see how these elements of positionality are played out in the research's conceptualisation and the methodological complexities they subsequently posed.

Methodological framework

My approach to this study has been influenced by feminism and poststructuralism. Post-structuralism incorporates a scepticism of 'scientific rationality', 'objective truth', 'neutrality' and 'meta-narratives' which provide seamless explanations of complex social phenomena. It is a product of what Lecourt (1975) has characterised as 'the decline of the absolutes' (p. 187). Rather than suggesting that research makes the world objectively and consciously knowable, post-structuralists suggest that knowledge is 'socially constituted, historically embedded, and valuationally based' (Lather, 1991, p. 52). Feminists who have used post-structuralism as a tool for making sense of their knowledge claims grapple with questions around the validity of this knowledge and its relationship to material realities (Lather, 1991; Jones, 1992; Middleton, 1995). One of the dilemmas feminists employing post-structural tools face is wanting to understand and transform unjust gender relations while realising that the realities of gendered lives cannot be accessed directly nor can there be any general rules for deciding between competing accounts of them (Ramazanoglu and Holland, 2002, p. 105). In attempting to address some of these problems feminists have sought to become reflexive researchers. As Ramazanoglu and Holland (2002) explain:

> Reflexivity generally means attempting to make explicit the power relations and the exercise of power in the research process. It covers varying attempts tounpack what knowledge is contingent upon, how the researcher is socially situated, and how the research agenda/ process has been constituted.   (p. 118)

In this way feminist researchers highlight how the knowledge their research produces constitutes a partial and situated account. What is known, is always perceived from a context affecting the perception of what is to be known. In addition, this context is saturated by power relations that mean particular subjectivities offer points of 'ruling' (Walkerdine, 1984). In the next section, I explore how these points of ruling affected the knowledge generated by this study as I was variously produced in the field work experience as 'insider' and 'outsider'.

The post-structural underpinnings of this research have also influenced how I have understood the process of data analysis. Although I have talked about the importance of understanding young people's own understandings of their sexual knowledge, subjectivities and practices this does not imply as Frosh, Phoenix and Pattman (2002) explain 'an uncritical acceptance of....[young people's]....versions of themselves' (Frosh, Phoenix, and Pattman, 2002, p. 4 my insertion). Instead these researchers describe how in their study about young masculinities:

> The theoretical frame in which much of this work is cast is one in which 'experience' itself is made problematic; that is, it is assumed

that all descriptions of experience are themselves 'discursive constructions', ways of making sense of things, of articulating specific
versions of self, identity and the world.   (p. 50)

I take this approach in my analysis of the qualitative data collected
during this research (see designing the methods and analysing the data
section). Young people's talk can be seen as representing the plethora
of discourses which circulate in any historical moment, regional location and cultural context. How young people negotiate these is dependent upon the discursive resources available to them and their
investments in the subject positions they offer. Like Willig (2001),
I understand discourses and language to constitute meaning (as opposed to simply describing it) and therefore as having the capacity to
constrain and enable behaviour and thought. When participants talk
about their sexual subjectivities, knowledge and practices I understand
them as engaging in particular constructions of themselves which
may vary as they move across different interactive settings (e.g. focus
groups, individual and couple interviews). In common with much discursive analysis, what is important to me is not determining which
narratives disclose the 'truth' about these young people, but how the
variables within a particular context produce such constructions. For
this reason throughout the book in the analysis of young people's narratives, attention is drawn to features of context, such as group composition in focus groups. The use of post-structuralism to inform this
analysis also assumes that the accounts young people present are
partial and that they may or may not be commensurate with material
reality. However, because discursive constructions enable and constrain
thought and practice I expect that the narratives young people offer
have some implications for their sexual practice.

The issue of what constitutes a feminist methodology has evoked
considerable debate amongst feminists with early conceptualisations
rendering this research for women, about women, by women (Bowles
and Klein, 1983, Roberts, 1981). One of the difficulties with this definition is that it presumes easy answers to complex ontological questions
about who is a woman and what makes her such. Some have claimed
that the defining point of feminist research is the reflexive way in
which the researcher locates themselves in the research process
(Harding, 1993; Stanley, 1992). Although, this reflexivity can 'also [be]
characteristic of other radical approaches to social research' such as
anti-racist and disability studies (Ramazanoglu and Holland, 2002,
p. 16). Ramazanoglu and Holland maintain that what is distinctive

about feminist research is 'the particular positioning of theory, epistemology, and ethics that enables feminist researchers to question "truths" and explore relations between knowledge and power' (Ramazanoglu and Holland, 2002, p. 16). This definition equates with Letherby's (2003) where feminist research is differentiated by 'the questions feminists ask, the location of the researcher within the process of research and within theorizing, and the intended purpose of the work produced' (p. 5).

The current research falls within the parameters of the above definition through its commitment to feminist politics and its exploration of the relationship between knowledge and power. Achieving gender equality within sexual relationships, where both partners exercise power over their own sexual decisions and behaviours and are free from physical/mental coercion and violence has been a motivating force for this work. Its sentiment reminds me of a comment made by Flax about feminist researchers who harbour 'enlightenment dreams' in their desire for emancipation, social progress and justice (Flax, 1992). The difficulty with these goals is that they draw on liberal humanist language that implies a common humanity and shared interests/connections between women that do not (always/often) exist. Such feminist politics sit in tension with the way analysis in this research has welcomed and incorporated a post-modern call for the disruption of meta narratives. This conflict remains unresolved for me and other feminists, who attempt to deal with it by interrogating their own constitution as knowing subjects with particular ethical positions and political identities (Ramazanoglu and Holland, 2002).

An increasingly common phenomenon in research labelled feminist is a shift away from an exclusive focus on women and women's issues to the incorporation of male participants and theories of masculinities (Francis and Skelton, 2001; Kelly, Burton, and Regan, 1994). The inclusion of males in feminist research has invited parallels in discussion to the move in universities from Women's Studies to Gender Studies. Some have perceived this as diluting feminist politics which has worked hard to carve spaces for women in which their interests and needs are prioritised. Given such arguments, the inclusion of 183 young men in this study might be considered to constitute a depoliticisation of the research's feminist aims. The decision to include young men was primarily underpinned by ontological assumptions about power which I perceive as enhancing the project's feminist objectives. Weedon (1987) explains that 'power is a relation' which inheres in difference and 'is a dynamic of control and lack of control' between discourses and the subjects they constitute (p. 113). To understand gendered power, requires an investigation of both points of reference of this power in/relation. If power 'inheres in difference' and through language masculinity and femininity are constituted as binary opposites, then masculinities and femininities are rendered inherently relational by their difference. Given that the nature of power is relational and the meaning of masculinity and femininity is derived from their inextricability in language and the difference this entails, I felt that an appropriate way to make sense of gendered power and (hetero)sexual subjectivities was to include young women *and* men. Subsequently, while the sample is weighted towards young women approximately a third is comprised of male volunteers.

Accessing a sample for research on sexuality

> Research on sexuality in schools is a complicated business. Linking 'sex' and 'school' together as a focus for study brings the researcher into direct contact with many of the symbolic boundaries that shape contemporary schooling. Constructions such as public/ private, adult/child, teacher/pupil, male/female, proper/improper organise social relations within the school in ways that seek to demarcate and prescribe the domain of the sexual. (Kehily, 2002, p. 5)

Gaining access to schools for research is often challenging due to a highly regulatory environment where finding adequate periods of time to engage students is difficult. Schools have increasingly congested timetables where academic pursuits take priority and the sound of the bell demands movement to the next class/activity. As Kehily (2002) articulates above, these difficulties are often compounded when it is research on sexuality that is being conducted. This is because some schools see sexuality issues as 'controversial' and feel any concentration on them may draw unwanted attention to the school and incite criticism from caregivers. It is these sorts of considerations which render studies of sexuality 'sensitive' because they 'raise questions about the kinds of research regarded as permissible in society and the extent to which research may encroach upon people's lives' (Lee, 1993, p. 1). In an attempt to circumvent some of these obstacles, schools in this study were pre-selected for their predisposition to support the research based upon what I knew about their health programmes and the teachers who ran them. Ideally I had hoped for a group of schools exhibiting a diverse range of characteristics and student compositions. For instance, an equal number of coeducation and single sex, lower and middle socio-economic status, urban/rural schools with alternative and traditional pedagogies. However, when participation is voluntary it is much harder to engineer a diverse and representative sample. Eventually, I secured participation from seven schools, one of which was single sex, the rest co-educational and accorded various decile ratings[2] by the Ministry of Education. Two of these schools had a reputation for being 'alternative' in their approach to students, curriculum, and administrative issues.

While 60 per cent of the sample were derived from secondary schools located predominately in one North Island city, approximately a third came from educational training programmes within the community. By including these young people I hoped to access a diverse array of subjects as many had left education early and had experienced different life trajectories to the school based group. Those eligible to attend this community based training had either fewer than three school certificate passes[3] and no qualification higher than sixth form certificate, were long term unemployed, refugees, or had been identified by the New Zealand Employment Service as requiring training for a number of reasons. These programmes were designed to assist trainees towards further training, employment and recognised qualifications, or credit towards them. Each programme specialised in a specific kind of training to match employment needs within industry, such as agriculture/horticulture, business office and computer skills, outdoor recreation, tourism and hospitality, engineering and metal trades as well as community and child care. To increase the likelihood of young people with different backgrounds, interests and perspectives participating I contacted programmes from 18 training fields that were situated in various localities (such as low socio-economic, middle class, urban/rural areas). Despite anticipating different responses from young people in the 'at school' and 'not at school' groups, answers were remarkably congruous rendering my intention to base data analysis on group differences difficult. To illustrate their similarity I have included the 'at school' (AS) and 'not at school' (NAS) descriptors as part of subject identification in the text.

In New Zealand one of the ways in which the sensitivities of research is managed is through a stringent ethics committee approval process. At the time of conducting this research I was enrolled as a doctoral student at a British university where I designed the project in my first year. The fieldwork took place in year two in New Zealand and in order to maximise my time there I acquired ethics approval from the university's ethics committee before leaving England. As this body were not an ethics committee approved by my funding organisation (The New Zealand Health Research Council), I also had to gain approval from a New Zealand based ethics committee. This meant struggling to meet the often conflicting regulations of two ethics committees, although the fact that the research had withstood the rigor of both added kudos for some organisations I contacted during the recruitment process. An underlying principal of both committees' regulations was the notion of informed consent. This meant providing on first contact with a school or community training programme an explanation of who I was, the research objectives and what participation involved. Typically this first contact took the form of a letter accompanied by a participant information sheet providing details about what possible participation would involve. In the case of schools, also enclosed was a Board of Trustees[4] consent form asking for signed permission to conduct the research and a letter endorsing the research and my professional conduct from my New Zealand supervisor. All forms had been checked and approved by both ethics committees before distribution.

The New Zealand based Ethics Committee were particularly concerned that given the research's focus on sexuality that those who took part were volunteers. This stipulation was premised upon the idea that as issues of sexuality may offend, embarrass or make young people feel uncomfortable they should not be compelled to participate. After recruiting for participants in schools, community training programmes and through advertisements placed in locations young people frequent (such as youth centres) the sample consisted of 515 volunteers aged 17–19 years old. This age range was selected because it had not previously been a focus of qualitative sexuality research in New Zealand and signalled a transitional stage in young people's lives from school to work or higher education. It is likely that because those who took part were volunteers that they shared particular characteristics which differentiated them from those that did not. Describing participation patterns in sexuality research Dunne, Martin, Bailey, Heath, Bucholz, Madden, and Statham (1997) argue that volunteers for these studies are likely to have 'higher levels of education, less conservative political sexual beliefs, and be less likely to attend church' (p. 851). If nothing else, participating in methods which involved speaking to a researcher and their contemporaries about sexuality, required feeling somewhat comfortable (if not curious) about this area.

Despite my (and the ethics committees') efforts to ensure participation was voluntary the notion of student's voluntary consent was problematic. By way of example, I would often be introduced to a class by a teacher who although indicating that participation in the study was voluntary would endorse the research as important and encourage students to participate (for more details see Allen, in press). For students who were eager to maintain the teacher's favour, participation often appeared to be less of a choice and more of a duty. In addition, sometimes when students expressed their desire not to participate they would be met by the teachers' warning that 'if you aren't involved in the research you have to do maths', a prospect many did not appear to relish. In fact on several occasions it was evident that class time had been allocated to participation in the research and no alternative activities arranged for those who preferred not to take part. The teacher's endorsement of the research as someone with authority and/or the offer of less attractive activities/or absence of them, meant that student consent was sometimes by default rather than an active decision.

Another consequence of recruiting a voluntary sample in school contexts, was an extremely low number of young people identifying themselves as gay, lesbian, bisexual or 'not sure' about their sexual orientation. Only two participants identified as same-sex attracted in the anonymous questionnaire and none overtly indicated this during the focus groups (for description of methods see below). Encountering a similar situation in their research on young men Frosh, Phoenix and Pattman (2002) suggest this is hardly surprising 'given "the presumption of heterosexuality" and the disgust and abhorrence which boys expressed when this was challenged, that no boys in these studies identified as gay' (p. 28). There is a burgeoning literature which reveals that gay, lesbian and bisexual identities and interests are silenced within schooling curricula and culture (Vincent and Ballard, 1997; Epstein and Johnson, 1998; Hillier, Dempsey, Harrison, Beale, Matthews, and Rosenthal 1998; Quinlivan and Town, 1999). Subsequently, schools are often not 'safe spaces' in which young people can divulge their sexual feelings without fear of discrimination or abuse. Research conducted in this setting can therefore inhibit students from self-identifying as same-sex attracted and given the possible dangers of this action it is arguable that we would even want them to. The low numbers of those identifying as same-sex attracted and my realisation that as a doctoral student with limited time, networks and fiscal means I was not materially or experientially equipped to recruit these participants, meant the research was confined to participants who described themselves as (hetero)sexual. I have always considered this aspect of the research unsatisfactory and have written about it in the course of a new project with a sample of gay, lesbian, bisexual, transgendered youth (Allen, in press A). However, I do not see this focus on (hetero)sexualities as precluding discussion of how the findings might inform sexuality education in ways that have positive effects for same-sex attracted and transgendered youth. This discussion takes place in Chapter 7.

Designing the methods and analysing data

In order to recognise the contextually specific nature of young people's narratives and capture their complexity, multiplicity and contradictions, the research employed an exploratory multi-method design. This comprised both qualitative and quantitative methods, which were used in a building-block fashion, each informing the design of the next. Focus groups were undertaken first to enable an initial familiarisation with young people's language around sexuality and the identification of issues they deemed important. This information was then incorporated within the framing of survey questions which aimed to collect contextual information about young people's sexual subjectivities, knowledge and practices. Themes which emerged from open ended survey questions were then used to identify possible topics for discussion within couple and individual interviews. The advantage of this sequencing was that it enabled familiarity with young people's language, perceptions and culture around sexual issues providing insights which could be used to shape ensuing methods.

Eighteen focus groups were conducted and included mixed gender or single gender groupings depending upon the preference of the young people participating. I had specified to those who facilitated my access to subjects that young people taking part should be friends to engender greater comfort and ease when talking about sexuality issues. Sessions typically lasted for an hour and were conducted in a variety of settings depending upon available space, which in schools usually meant an unoccupied class room. Community training programme focus groups took place in more diverse settings with one session held in a sports club, two in lunchrooms, several in conference rooms and another at the back of a restaurant. Discussion format was structured in two parts, the first around a series of loosely structured open-ended questions about young people's (hetero)sexual relationships. The purpose of these questions was to obtain a sense of how participants perceived (hetero)sexual relationships, what was important about them, what problems they experienced and how they constructed and articulated these in a group setting. In the second part, an activity involving media images depicting dominant and alternative discourses of (hetero)sexuality was undertaken.[5] Offering both typical and alternative representations of relationships was an attempt to disrupt conventional meanings which traditionally cohere around (hetero)-sexuality. Participants were asked to select an image they wanted to talk about and indicated whether the messages it communicated about relationships resonated with what they knew about them. These pictures provided a framework for looking at sexuality where participants could juxtapose their personal experiences with media constructions.

The second method involved the completion of 411 anonymous questionnaires designed to economically gather information that could contextualise data acquired from the qualitative methods. This included information about the average length of relationships, the age at which young people started dating, how many were sexually active and what kind of sexual knowledge they perceived they had. The questionnaire was divided into three sections mirroring the research's central areas of interests; young people's sexual knowledge, subjectivities and practices. Sexual knowledge questions concentrated on aspects of knowledge not traditionally explored in sexuality education such as; whether young people knew how to get what they wanted from a sexual relationship; sexual positions and techniques; as well as knowledge about how to avoid unwanted sexual activity. The focus of the sexual subjectivity section was how young people perceived themselves as sexual subjects and their perception of their sexual bodies as an extension of this. Questions on young people's practice sought to uncover how they negotiated their relationships, how they communicated their desires to a partner and, what their experiences of relationships were. Unlike the other methods involving an audience of either myself or the peer group, the anonymity of the questionnaire offered a sense of privacy. Including this type of method was an attempt to overcome some young people's embarrassment or fear in expressing thoughts about sexuality in the presence of others.

As the research sought to understand young people's perceptions of their (hetero)sexual practice it incorporated an examination of the couple context. The purpose here was to reflect on how partners interacted together and how they portrayed their sexual decision making as a couple. Due to strong cultural taboos surrounding direct observation of decision making during sexual activity another method for studying this had to be devised. I settled on a combination of firstly, an activity which the couple participated in together, followed by an interview conducted with each partner separately. These were facilitated by me and took place in settings where couples indicated they would feel comfortable, either in their own homes, my home or a meeting room in the university where I was teaching at the time. As with the focus groups, sessions were tape recorded with the prior written permission of those participating. Interview length ranged between 2–3.5 hours, usually dictated by whether other people required the space we were occupying.

In all, six couples drawn from both the 'at school' and 'not at school' samples participated (Appendix). They included individuals who identified as Pakeha, Maori and Samoan and whose relationship length varied from between three months to over three years. The activity involved couples sorting cards with a series of phrases about their relationship, sexual subjectivities and knowledge into piles under three headings: 'often happens or happened in our relationship', 'sometimes happens or happened in our relationship' and 'never happens in our relationship'. The card phrases centred around issues identified by young people participating in the focus groups as points of contention in relationships such as, condom use or one partner not wanting to engage in sexual activity or particular kinds of sexual activity. The activity's main function was to see where couples placed cards, an action which offered insight into their construction of negotiation and decision making around sexual issues in their relationship. Interviewing each partner individually following the activity aimed to provide an opportunity to revisit what they had said or had done in the activity context. This allowed them to explain why they agreed or disagreed with their partner, in a context where they were uninhibited by that person's presence. It also enabled exploration of those issues, which arose in the activity that were not fully examined, or which I wanted to understand more comprehensively. Through comparison with the couple activity transcripts, the individual interview offered an analysis of how subject's constitution of their subjectivity was modified in the presence of the other partner.

A last element of this method was the completion of a 'Pleasure Sheet'. This checklist was designed to obtain information about what young people found physically pleasurable about sexual activity (such as kissing, oral and anal sex, masturbation etc.) and desirable in terms of relationship satisfaction (security, trust, support, honesty, sexual satisfaction etc.). It aimed to provide data about young people's corporeal experiences of desire and pleasure a topic which is typically missing from sexuality education (Fine, 1988). While one partner was with me, the other completed a two page checklist, the first asking what they desired from their current relationship and whether these desires were being met. The next page enquired about types of sexual activity they had currently and previously engaged in and whether these were experienced as pleasurable. How these data might contribute to the constitution of what I term 'a discourse of erotics' in sexuality education is examined in Chapter 6.

A concern expressed by one of the ethics committees about interviewing couples was the potential for conflict between partners. While I was unaware of any adverse effects regarding the couple's relationship, two participants commented on its beneficial impact. One young woman Ngaire who participated with her partner of 9.5 months, explained that she felt flattered he had agreed to take part and saw this as indicating his deep felt feelings for her. Responding to a question about the possible effects the research may have on her relationship Ngaire explained:

> ....I did learn that uhm.... oh the way he feels about me... Probably because uhm he's offered to do this interview. But I've just found it amazing that he's actually said you know oh I really care about her and things like that. Cause...He has said it to me but never to anyone around us you know.
>
> (II, NAS, 19)

Becky another of the interviewees expressed similar sentiments viewing her partner's involvement as a sign of renewed commitment to, and respect for her following a break in their relationship. Instead of tension and conflict posing difficulties for the research, participants were eager to represent their relationship as 'ideal' and 'difficulty free'. This phenomenon was highlighted in the individual interview when I enquired about how the couple activity had proceeded. Tim who had been going out with his girlfriend Emma for six months revealed how he had been wary of saying anything which would cause conflict.

> Tim: ....yeah my main insecurity was saying stuff when Emma was here you know. Just in case we had a big disagreement *(laugh)*. If I said let's put it here and she said let's put it here [the activity card] and it would be like ah oh *(laugh)* time for that relationship compromise!
>
> (II, NAS, 19)

Similarly, Chris who had been seeing Cam for three months explained how he had steered away from particular subjects in order to avoid uncomfortable confrontations.

> Louisa: So how did you find the couple activity? Was it difficult to talk about or......?
> Chris: Ah no it wasn't because it was talking about Cam and yeah. I would have been more uncomfortable talking about other experiences. Uhmm....
> Louisa: Other previous experiences that you've had before Cam?
> Chris: Yeah.

(II, NAS, 19)

Cam had also monitored her talk about previous partners explaining in the individual interview that; 'It's kind of hard to say things like that [about previous boyfriends]. I find it hard to say like...things like that to him because he's very sensitive about stuff like that anyway...'. An advantage of having multiple methods was that it provided opportunities for these sorts of tensions to be revealed and indicated how different narratives were produced in various research contexts. The implications of these tensions for understanding how young people constituted their relationship practice are explored in Chapter 6.

As the study utilised both quantitative and qualitative methods, data analysis involved a combination of different techniques. Due to degree regulations which stipulated no part of the research could be undertaken by anyone else, I transcribed the 35 tapes generated by the focus groups, individual interviews and couple activity myself. By doing this I acquired a strong sense of themes emerging from each transcript and was able to use these as an early basis for data classification. These themes were then ordered under the main research concepts of 'sexual knowledge', 'sexual subjectivity' and 'sexual practice'. As gender was a structural component of my analysis I searched within these headings for gender differences in young people's talk. Next I explored the points at which young people's narratives about for example 'knowledge and practice' or 'subjectivity and knowledge' intersected. In this way the relationships young people made between such concepts and how these might pertain to the notion of a 'gap' could be ascertained (a gender analysis was also performed on these themes). The multiple method approach allowed for a kind of funnel analysis where general themes which had emerged in focus groups could be studied as they surfaced in the more intimate context of individual and couple interviews. For instance, young women's talk about dissatisfaction with their bodies could be analysed and contextualised in terms of the personal experience of an individual interview participant. The final stage of analysis compared findings on the main themes with existing theory and research in the field to determine similarities and differences as well as the current work's possible contribution.

The quantitative data which emerged from the questionnaire necessitated a different analytical approach. To handle the volume of figures I had acquired I employed the software package SPSS.[6] As the aim of the quantitative data was to provide general patterns about young people's sexual knowledge practices and subjectivities I used standard statistical measures. These included chi-squares and T-tests to establish gender and 'not at school'/'at school' group differences. I then read the questionnaire data in relation to the qualitative material and used them to explain or illustrate findings from each of the methods. Depending on the nature of open-ended survey questions I looked for patterns and anomalies in answers across the sample and particularly in relation to the category of gender.

Through my 'I' : The researcher/researched relationship

Fine (1994) argues that researchers need to 'work the hyphen between Self and Other'. By this she means acknowledging and examining the researcher/researched relationship in terms of the multiple subjectivities and contexts in which we are constituted. Reworking the hyphen challenges traditional research objectives of neutrality and authoritative authorship by asking the researched/researcher to consider:

> .....what is, and is not, 'happening between', within the negotiated relations of whose story is being told, why, to whom, with what interpretation, and whose story is being shadowed, why, for whom, and with what consequence (Fine 1994, p. 72).

Engaging in this involves exploring positionality, or the social position of the knower and recognising this as both fluid and contextual (Rhoads, 1997). In her paper on cross cultural research with Pasifika students, Petelo (1997) argues that we examine our positionality by asking ourselves the following questions: 'How do I construct others? How do participants position me? How do I see participants positioning themselves? How would I like to be positioned by the participants?' (p. 6). There are many aspects of my positionality (being young as

I was 24 when conducting the fieldwork, a woman, a university researcher, a feminist, middleclass, (hetero)sexual, Pakeha) that have framed interaction in this research and as I have indicated affected the knowledge produced. I choose here to foreground my identity as a Pakeha/Palagi[7] as this raised important methodological dilemmas for the research which highlight it as a partial and situated account.

As a Pakeha/Palagi I was 'outsider' or 'other' to members of Maori and Pasifika communities, a positioning first apparent in pre-field work discussions about what the study might offer these groups. Meetings with these communities were a valuable experience in reversing the relations of ruling, where I was constituted as lacking knowledge of Maori Kaupapa[8] appropriate to conducting the research. Many researchers have found themselves in this position when their own culture is other to that of their participants. As Fine and Weis (1996) note about their joint research project, 'a good number of people of colour don't trust two white women academics to do them or their communities much good' (p. 271). To understand this reaction in the New Zealand context it is important to document historical relations between Maori and Pakeha. These can be conceptualised as colonisation of Maori not only in terms of land, resources, language and culture, but also of the mind (Simon, 1994). The failure of the Crown to honour 'te tiriti o Waitangi' (The Treaty of Waitangi), and provide Maori with sovereignty (rangatiratanga) as envisaged by them, has forced Maori to face as Walker (1990) has coined it 'a struggle without end'[9]. This struggle extends to the realms of academic research where Maori have had to 'listen to Pakeha define their problems and prescribe their solutions' (Smith, 1992). There are too many examples of ethnocentric research by Pakeha that portray Maori as some how deficient, difficult, delinquent and helpless victims (Smith, 1990). Cram (1997) explains that 'many Pakeha researchers have built their careers on the backs of Maori – their research has satisfied the criteria set by Pakeha institutions but offered nothing back to the Maori community in return' (p. 45). When such studies are not designed in accordance with Maori kaupapa or 'culturally intelligible and acceptable frames of reference' (Stewart and Williams cited in Cram 1997, p. 48) they become research **on** Maori rather than **for** Maori (Smith, 1990).

Senior members of the community I consulted had previously been subjected to such damaging research and were not going to risk any form of neo-colonisation. While two young people from the community centre I visited were eager and permitted to fill out my questionnaire, the centre director implied their involvement in an interview would not be supported. As this was a pre-fieldwork visit I did not use a tape recorder or ask the centre director's permission to use her words so I cannot quote her. Instead I offer these notes written up in the afternoon after my visit.

> *Today was my first 'real' field work experience to consult members of Maori and Pacific Island communities about the research I'm proposing. When I rang up they didn't sound too interested in the study or me coming, but as they gave me a time to show up, I went along. When I first got there I was taken into the kitchen and sat around a table with 4 young people. As I started to explain about the research I became quite nervous in unfamiliar surroundings and acutely aware of how Pakeha I sounded. The centre director was clued up. She asked all sorts of things about the research and was especially concerned with the questions about ethnicity in the questionnaire and how this particular data would be used in my analysis. Her other concern was that as a Pakeha researcher I would be working with Maori and Pasifika youth and yet I was not a member of these communities (this is a concern I share). She said she would prefer Maori to do a study on Maori and I agreed with her that this would be ideal. I explained how as a Ph.D student I didn't have the resources to have a team of Maori and Pasifika researchers.*

The following week I visited a community group that worked with Pasifika youth and its manager expressed more hesitant reservations about my doing the research. These concerns cohered around the way Pasifika youth might not take me seriously because I did not command the respect of an elder. Like Maori members of the group above, the director was concerned that my Palagi worldview would force an interpretation of these data that would be ethnocentric. The manager then introduced me to some of the young people employed by the centre and I asked them if they thought my being Palagi would raise any difficulties in conducting the research. Their response was 'we don't see any problem with it'. When I explained the reaction I had received from senior members of their communities, and asked why they thought they had said this they replied, 'I don't know, they're just protective'.

For me, the outcome of these discussions was a feeling of immobility. This was produced by a fragmentation of subjectivity that simultaneously cast me as a member of the ethnic majority and thus exercising power, but also disempowered by/within my own research process. Bhopal (1997) explains the fragmentation of subjectivity as due to complicated interrelations of difference and 'othering' when researchers and the researched are positioned simultaneously in dominant and subordinate positions in relation to varied 'selves' and 'others'. This feeling of disempowerment stemmed from a more complicated source than being told by community members that it was inappropriate to conduct research with Maori and Pasifika youth. I was aware of this before embarking on the project. Rather it emerged from my realisation and subsequent frustration, at not being able to escape the exercise of power my cultural gaze entailed and was compounded by my inexperience as a social researcher.

To make sense of these feelings of immobility I find it useful to draw on Jones' (1999) work concerning *hearing* the voices of oppressed or marginalised groups. Jones employs Barthe's ideas about the power of the reader to make the text in explaining the way in which dominant group members (in a classroom context) do not have the 'ears to hear' marginalised groups. Jones says,

> ......that understanding lies not in (a text's) origin but in its destination then it is not so much the subaltern's 'voice' but its 'heard-voice', its audience in the classroom, which becomes the key player in meaning. When dominant group members are unable to understand the speaker because they do not have the 'ears to hear', then the subaltern's desire or ability to speak in the classroom must be reduced dramatically.   (p. 308)

Although Jones is talking about cross cultural dialogue in the classroom I find her thoughts useful for interpreting my own response and that of the participants above. From this perspective their refusal might be seen as a recognition that 'I did not have the ears to hear' what they deemed important for the research. As a non-Maori and non-Pasific person I could not hear or make sense of their meanings in culturally appropriate ways and therefore participating in the research was fruitless for them. My sense of unease (expressed in my field notes), can be attributed to a realisation that I did not/could not have access to this knowledge. I could not know, because 'I didn't have the ears to hear' not for want of them, but for lack of them. My ethnocentricity meant that any attempt to know would involve another kind of colonisation. My reaction to this 'not being able to know' was an academic paralysis and feeling of impotency to address the situation.

This feeling of immobilisation was heightened by the institutional power of the university in which I was enrolled which set degree criteria and thus the possibilities for research design. A potential way of overcoming the study's lack of kaupapa was to establish a research partnership acceptable to each ethnic community. However, regulations for my degree dictated that all research be undertaken by candidates themselves. As a doctoral student I felt I exercised minimal power to change this situation, and felt additional pressure not to sabotage the attainment of a degree for which the Health Research Council had invested funding. The rigidity of these rules were palpable in the declaration at the front of my thesis which I was required to sign on the day of submission. This read, 'the data, analysis and conclusions are the result of my own work and include nothing which is the outcome of work done in collaboration'.

My ethnic otherness and the institutional constraints I experienced significantly affected the knowledge this research could produce. My feelings of academic impotency found expression in what I see as unsatisfactory methodological decisions. I characterise these as emerging from academic impotency because my sense of my own agency felt so limited. One of these decisions was that while I accepted involvement from volunteers of any ethnic group, I did not use ethnicity as a category of analysis. At the time, this decision rested on several bases. The first was my desire to avoid (re)producing accounts of Maori or Pasifika youth that evoked a deficit model as expressed in the fears of the centre managers above. Any such analysis would enable a comparison with Pakeha/Palagi and a reading of the data that could potentially constitute Maori and Pasifika youth as lacking (if not by me, then by others).

Secondly, 'without the ears to hear' I believed that I could not conduct and analyse the research through anything but a Pakeha frame of understanding. This inability was highlighted during focus groups where youth were predominantly Maori and Pasifika. The group banter pertaining specifically to cultural knowledge was inaccessible to me, and because of this I felt I lost subtleties of information about (hetero)-sexual relationships which it offered. Not always being able to join in with the laughter and share in the obvious pleasure it elicited revealed the 'limits of my ears'. As I believed any kind of representation would involve an objectification that would render the research **on**, not **for** Maori and Pasifika peoples it was inappropriate to 'speak on their behalf'. At the same time however, I did not want to disappoint or exclude any young person who had expressed interest in participating in the research so I accepted participation from all who volunteered. One way around this might have been to accept participation from all ethnic groups and then filter data so that only those of Pakeha remained. However this left me with the predicament of how to delineate Pakeha identity and how would I categorise participants who identified themselves as both Pakeha and Maori?

This methodological decision was highly problematic in that paradoxically my inclusion of Maori and Pasifika youth as subjects in the research rendered them invisible. Without identifying ethnicity as a category of analysis I had silenced this element of young people's subjectivities and its effect upon their understanding of themselves as sexual people. In negating any ethnic analysis, I had also reinforced the status quo in which the voices of Pakeha are heard and thus privileged. When I originally made this decision, it had been my intention to offer data collected on particular ethnic groups to their communities. I had hoped that doing this would in some way give voice to ethnicity and some control to Maori and Pacific groups in determining it. Unsurprisingly, my offers of these data to key people within these communities were not welcomed. In my naivety I was trying to 'fit' Maori and Pasifika youth into a research project which had not involved them from its inception. As the research question had found shape during the first year of my degree in Britain and the methods developed before returning to New Zealand, establishing working relationships with these communities had been logistically impossible. This meant that the research did not reflect their 'kaupapa' or offer them reason for investment in it. As a consequence this research cannot offer diverse cultural insights into young people's experience of sexuality. This work remains to be undertaken by the diverse ethnic communities of which New Zealand society is now comprised.

3
Sperm Meets Egg?
Young People's Conceptualisations
of Sexual Knowledge

In this chapter I argue that if we are to better comprehend why some young people do not put their knowledge into practice we first need to establish how young people themselves conceptualise 'sexual knowledge'. Instead of examining the way in which sexuality education has constituted sexual knowledge, this research sought to uncover young people's own definition of it. Starting from this premise means recognising young people have agency as sexual subjects and that their own view points are important. From this perspective young people are not simply empty vessels who absorb the knowledge to which sexuality education exposes them, but instead actively engage with the information they receive. This chapter then, prioritises young people's own understanding of the supposed knowledge/practice 'gap' and calls conventional theorisations of this phenomenon into question.

As part of this exploration, sources of sexual knowledge are also examined. Knowing where young people gain access to information about sexual matters provides insights into the kind of knowledge they seek and deem authoritative. One of the overall purposes of this book is to tease out the relationships between sexual knowledge, subjectivities and practices as a way of thinking through the 'gap' phenomenon. This chapter concentrates on young people's view of the relationship between sexual knowledge and practice and involves analysing what if any impact they perceive their sexual knowledge to have on sexual practice. I aim to demonstrate that understanding how young people think about sexual knowledge is imperative for formulating sexuality education programmes that satisfy their needs and interests.

35

The constitution of sexual knowledge within sexuality education programmes

Before plunging the reader into the research findings it is necessary to establish how young people's sexual knowledge has traditionally been constituted within the context of sexuality education. This particular conceptualisation of sexual knowledge is inherent within the idea of 'a gap phenomenon' and therefore necessary to any examination of why this 'gap' might occur. It also provides a backdrop to thinking about young people's own understandings of sexual knowledge and how these may differ from traditional conceptualisations.

Sexual knowledge has a particular constitution within sexuality education programmes. Such conceptualisations are tied to sex education's philosophical underpinnings where the knowledge it offers is deemed necessary for students' (physical) well-being. Through the presentation of information about bodies and sexuality this knowledge serves as a disciplinary technique. The disciplinary nature of sexuality education's constitution of sexual knowledge can be described as follows:

> Sex Education is the formal expression of the training and disciplin-
> ing of bodies in this most crucial arena, for Sex Education both con-
> structs and confirms the categories of 'normal' and 'deviant' that are
> central to the regulation of social life (Zita cited in Thorogood,
> 2000, p. 429).

Historically, in countries like New Zealand, Britain, Australia and the US this disciplining of bodies has centred around the curtailment of 'social problems' such as illegitimacy, sexually transmitted infections, teenage pregnancies and abortions. These have been constituted as undesirable consequences of sexual activity because of the social and economic pressures they place on the state. Sexuality education has provided a vehicle for arming young people with the 'right knowledge' to avert such 'disasters' leading some commentators to assert that, 'the dispensing of sexual knowledge as a prophylactic for the unwelcome consequences of free-wheeling sexual behaviour is the cornerstone of modern sexuality education' (Sears, 1992, p. 17). This agenda means that sexuality education's conceptualisation of knowledge has often been limited to facts that are aligned to the prevention of unwanted outcomes. For instance, a concentration on the range and pathology of sexually transmissible infections.

Some might argue that this is an outdated portrayal of sexuality education's objectives where greater emphasis now falls on the importance of healthy relationships, personal autonomy and the attainment of know-ledge and skills around physical and emotional development. Such shifts are reflected in participatory pedagogies which have largely replaced the formerly didactic methods of knowledge transmission. While the rhetoric has undergone modification a primary focus of much sexuality education remains the reduction of illegitimacy, sexually transmitted infections, teenage pregnancies and abortions. In Britain for instance, New Right dis-course reconstitutes talk about illegitimacy in terms of the importance of 'family' with the 1994 DfE Circular (para. 8) stating that 'Pupils should accordingly be encouraged to appreciate the value of stable family life, marriage and the responsibilities of parenthood'. In New Zealand, page one of the Ministry of Health's 'Sexual and Reproductive Health Strategy' establishes government concern in two key areas, 'New Zealand's increas-ing number of sexually transmitted infections' and 'the high level of unin-tended/ unwanted pregnancies'. These policy directives impact upon the knowledge transmitted in sexuality education so that issues which aid in the prevention of unplanned pregnancy and STIs (such as contraceptive and condom use) are given precedence.

These policy concerns are apparent in examples of contemporary health education curriculum content. An informal survey of work-book resources used in New Zealand secondary schools indicates that while more holistic concerns around young people's sexual health are included, greater emphasis is still placed on preventing 'negative' out-comes of sexual behaviour. Typically these booklets are compiled by those who teach health education and often comprise information from a diversity of sources (for example, the Ministry of Education and Family Planning). In one booklet used in an all boys school, 26 of its 118 pages relate to a section entitled 'Sex, Sexuality and Relationships'. While this indicates that sexuality issues are being addressed, all but four of these pages are devoted to sexually transmitted infections (labelled STDs with a special emphasis on HIV/AIDS), reproductive biology, personal hygiene and abortion statistics.

Sexual knowledge is constituted in this booklet through learning expectations and as evidenced in this selection of questions and exercises students are obliged to undertake:

- What might AIDS victims die from?
- The amount of blood lost in an average period is? a) 30–60 mls b) 60–90 mls c) 90–120 mls
- It is advisable to wash all over; a) Once a week b) Once a day c) Twice a day.

- Activity One: Match labels and functions of body parts with diagrammatical representations of female and male reproductive organs.
- Activity Two: Students should familiarise themselves with a matrix which lists sexually transmitted infections their symptoms and effects, accompanied by a drawing indicating 'Comparative Sizes of Human Sperm and Common STD Viruses'.

While it is not evident how comparing the size of sperm and STIs might contribute to young people's sexual well-being, its insertion signifies its perceived importance as 'knowledge'. As constituted by the above questions and activities sexual knowledge is equated with disease (STIs and specifically HIV), personal hygiene (when to wash) and reproductive biology (as seen in the naming of reproductive body parts and blood lost during menstruation). With a clinical focus on pathology and physiology this selection of sexual knowledge evokes negative associations through its reference to disease, bodily odour and blood loss. There is a noticeable absence of sexual knowledge as it is related to sensual corporeality, sexual diversity and personal empowerment with regard to people who are HIV positive as well as menstruating women.

Only four pages of this booklet traverse emotional and mental issues in sexual relationships. One such page provides commentaries on 'what is real love', assuring young people that 'being attracted to people the same sex as you is really ok, it doesn't in any way change who you are'. While attempting to be inclusive the effect of asserting that someone is still 'normal' *even if* they are gay renders them as 'other' through the need for such a remark. These statements are also undermined by the way they appear on the same page as a box which contains statistics about the seemingly unrelated issue of 'abortion'. Without supporting description, the unexplained presence of these statistics is a reminder to students that love and relationships are (hetero)sexual and can lead to unwanted consequences. This booklet epitomises many others currently in existence in New Zealand secondary schools and which researchers criticise as carrying a similar emphasis on negative and harmful elements of sexuality (Elliott, 2003).

Such examples remind us that despite calls for more comprehensive approaches to sexuality education the constitution of sexual knowledge in many curriculum resources is still intimately tied to preventing 'negative' outcomes of sexual activity. Given that governmental policy around sexual issues is framed and driven by the aim of reducing unwanted/unplanned pregnancies and sexually transmitted infections this is not surprising (Ministry of Health, 2001). The technocratic and rationalist principals underlying the operation of most schools mean that generally only knowledge perceived as supportive of these aims gets squeezed into the curriculum. Addressing the physical, emotional, mental and spiritual elements of young people's sexual well-being are not pursued in themselves, but in the hope that a by-product will be a reduction in 'negative' outcomes of sexual behaviour. Subsequently, the constitution of the sexually knowledgeable subject in these programmes is the student who can correctly answer questions about STIs and who knows the difference between the vas deferens and a gamete.

Young people's constitution of sexual knowledge

When researchers assert that a 'gap' exists between the knowledge acquired from sexuality education and its (consistent) translation into practice by some young people, they are referring to this form of sexual knowledge. It is 'official' knowledge derived within an institutional context like sexuality education, to fulfil a particular remit. The findings from this research suggest young people's own understandings of sexual knowledge diverge from sexuality education's constitution of it and this conceptualisation contains several nuances. This section reveals how this discrepancy points to a flaw in thinking which underpins the 'gap' phenomenon and highlights possible reasons for why sexuality education's messages are not always actioned.

The nuances inherent within young people's talk about sexual knowledge were apparent during the couple activity when participants were asked to decide whether it 'sometimes', 'often' or 'never' occurred in their relationship that 'one partner felt that the other knew more about sexual activity'. In the course of establishing who was most knowledgeable couples delineated their understandings of sexual knowledge. From these discussions it emerged that they were conceptualising sexual knowledge in two ways. The first, was a notion of sexual knowledge derived second hand from sources such as friends, sexuality education, television and magazines. The other was sexual knowledge attained first-hand from practical experience. While this distinction was made by all of the participants it was articulated most clearly by two of the couples who had each been together for over three years. When faced with the question of who was most knowledgeable Peter and Amy replied;

Peter: It depends on how...if you define that as in they know more as in they've done more, or they know more as in they've read lots about it or something like that. It depends.

Amy: I think that you probably had read a bit more than I had but, I'd been out with more people so I sort of was a bit more clued up about...

Peter: On what?

Amy: On the what and where sort of thing but not, not actually having done anything. I mean I knew how to snog and that which is what you didn't really have any idea about, but you'd read more than I had and definitely had more ideas *(laugh)* in that little head.

                                                                                  (CA, AS, mixed)

Ashby and Becky drew out a similar distinction between these conceptualisations of sexual knowledge in their discussion.

Ashby: I think she might know a bit more than me because she's been on a course and stuff *(laugh)*.

Becky: I know more practical stuff but only, not to the extent of positions and stuff *(laugh)* but to the extent of what's safe and what's not and how to protect myself in different situations and you know that's sort of what I know from like Family Planning and sexual health cause you know we went on a big hui[1] and they did all their different programmes.

                                                                                  (CA, AS, 17)

Both sets of couples separate sexual knowledge into the categories of practical experience 'as in they've done more' (Peter) , 'I knew how to snog' (Amy), 'to the extent of positions and stuff' (Amy) and knowledge gained from secondary sources such as books and a peer sexuality training course 'she's been on a course and stuff' (Ashby), 'I know more...to the extent of what's safe and what's not...cause you know we went on a big hui' (Becky). This latter conceptualisation of sexual knowledge entailed an intellectual rather than practical grasp of things sexual and was alluded to by other participants on a number of occasions.

Ngaire and George produced this notion of sexual knowledge when they talked about sexual information gleaned from friends.

Louisa: So where do you get your knowledge about sex from?
George: ...just like talking about it with friends and stuff like that.
Ngaire: Yeah I would too...I spoke about it at school with my friends.

(CA, NAS, mixed)

Similarly, in deciding who was most knowledgeable in the couple activity session, Chris explained in his individual interview how such knowledge was derived from secondary sources.

Chris: Like uhm I was sort of I always was interested in stuff. Like uhm I was sort of, I always watch the news and everything and I sort of knew of some sexual diseases but not others you know like uhm AIDS and uhm stuff like that but nothing much else.

(II, NAS, 19)

What is particularly interesting about such conceptualisations of sexual knowledge is that participants appeared to give greater weighting to knowledge gained from practical experience. This prioritisation became apparent when it was the partner with most practical experience who was invariably deemed most knowledgeable. The greater value attributed to sexual knowledge gained from practice was captured in statements used to qualify a couple's decision about who they had named as most knowledgeable. With reference to his girlfriend Cam's superior sexual knowledge Chris certified that 'Cam has had a lot more sexual relationships than I have' (CA, NAS, 19).

In Becky and Ashby's case it meant that despite Becky's considerable knowledge and experience as a peer sexuality educator she conceded to her boyfriend Ashby that:

Becky: I think that you might know a bit more than me because you've been with more people than me.

(CA, NAS, 17)

Both Nina and Neil agreed that Nina's Christian beliefs about refraining from premarital sex meant her sexual experience had been limited before meeting Neil and therefore he was the more sexually knowledgeable. Neil's promotion to most knowledgeable was despite the fact that Nina had received sexuality education at school while he hadn't.

Nina:　You knew that I didn't know anything [addressing Neil]. You knew that I'd only like kissed two people before. He knew I just hadn't done anything other than kissing and holding hands.

(CA, AS, 17)

This prioritisation of practical experience was also revealed in Tim and Emma's discussions about who was most sexually knowledgeable. Although Tim was sexually inexperienced he portrayed himself as 'imaginative' and sexually knowledgeable from a variety of secondary sources. He expressed agreement however when his girlfriend asserted 'I am probably a bit more experienced because I've had more partners' (CA, AS, 17).

It is likely that sexual knowledge gained from practical experience was more highly valued because of the kinds of discourses which inform young people's ideas about how learning is best achieved. Learning by doing is constituted as superior because it entails the 'real thing'. Actual engagement in an activity often offers a level of embodied knowledge not obtainable from secondary sources. A comparative example, might be learning by reading about how to assemble a car engine compared with the experience of undertaking this task. Someone who has gained sexual knowledge from first hand experience is therefore deemed to have greater insight into sexual matters than a student with no practical sexual experience but has been trained as a peer sexuality educator.

The greater value participants attributed to knowledge acquired through practice was not simply due to having 'done it', but also the type of sexual knowledge elicited through these experiences. First hand sexual knowledge is more likely to provide insights into corporeal sensuality (the subject's own and others), the logistics of bodily movement, emotional issues and the micro politics of sexual decision making. As indicated in the previous section this is less likely to be the kind of sexual knowledge conveyed by sexuality programmes and young people may view experience as a (pleasurable) way of discovering it. Using relationships as a means of finding out about things sexual was a motivation for some young people in the research as the following conversation revealed:

Louisa:　So why do you reckon young people get involved in boyfriend/girlfriend relationships?
Georgia:　To find things out.

Lita: Yeah to discover.
Sandra: Yeah.
Louisa: So when you say 'discover' what exactly do you mean?
Georgia: Physical.
April: Not much mental really.
Sandra: Yeah.
Georgia: Who cares about intelligence?
Sandra: Just use them to gain pleasure.

(FG, AS, mixed)

For these young women relationships clearly offered an opportunity to learn about physical intimacies and embodied sensations. There is a suggestion of a dual purpose of relationships here in terms of fulfilling their curiosity to know about their partner's body (as opposed to their personality in April's case) and to (re)discover their own corporeality through interaction with a male who they might 'use to gain pleasure' (Sandra). Young men in Kehily's (2001) study indicate a similar agenda in their descriptions of how (hetero)sexual relationships enable them to gain sexual knowledge through exploration of their girlfriend's bodies and talking with them.

In the absence of a curriculum which acknowledges and explores these aspects of embodied knowledge, relationship practice offers a (pleasurable) means by which such information can be acquired.

Knowledge obtained from sexual encounters may also hold more value because possessing it implies an esteemed sexual subjectivity, especially for young men. Acquiring sexual information in the context of hetero-sexual relationships offers young men access to knowledge that is 'highly prized and based on a "doing" that enhances masculine identities' (Kehily, 2001, p. 184). Masculinities are often articulated in terms of activity and performance so that gaining sexual knowledge through 'doing' provides young men with status. The relationship between sexual subjectivities, knowledge and their gendered nature is explored in Chapter 4. As the next section indicates, the means of acquisition of sexual information is not the only factor influencing young people's conceptualisation of sexual knowledge. In addition participants distinguished between different discourses of sexual knowledge demonstrating their greater interest in and preference for one.

Official discourses of sexual knowledge

Another nuance apparent in young people's conceptualisation of sexual knowledge was that this was comprised of two types of discourses 'official' and 'erotic'. The official discourse made reference to issues generally addressed within sexuality education such as knowledge about contraception, condom use and sexually transmissible infections. This discourse promoted individual responsibility, the importance of being organised (that is, prepared with condoms/contraception) and the need to exercise personal constraint (for example, don't let alcohol, drugs, desire cloud your judgement). This constitution of sexual knowledge was alluded to in Becky's comments above when she said she felt knowledgeable about '....what's safe and what's not and how to protect myself in different situations and you know that's sort of what I know from like Family Planning and sexual health'. Chris invoked a similar notion of sexual knowledge when he recounted '....I always watch the news and everything and I sort of knew of some sexual diseases but not others you know like uhm AIDS and uhm stuff like that....'

A sense of officialdom surrounds this discourse within which a moral voice makes appeal to clinical facts and scientific 'truths' about the need to act in sexually responsible ways. Chris's allusion to 'the news' as a source of his sexual information aligns such knowledge with a reputation for fact. Similarly, Becky's reference to Family Planning associates her knowledge with an organisation trusted to provide reliable, impartial and expert advice on sexual matters. As a consequence of its concern with safety this discourse denies passion and conflicts with ideas of sexual activity involving eroticism and spontaneity. In referring to this kind of sexual information young people imply that an 'official discourse' is a feature of their conceptualisation of sexual knowledge. This challenges a commonly held perception that young people are oblivious to or uninterested in, responsibility and the potentially negative consequences associated with sexual activity. The fact that this discourse was referred to less frequently and with less enthusiasm than a discourse of erotics, reveals another nuance in the way young people conceptualised their sexual knowledge.

Unofficial discourses of sexual knowledge: A discourse of erotics

A second discourse of sexual knowledge evident in young people's talk described the 'lived' experience of sexual activity and details of this interaction. This information was more personal centring on emotional and corporeal feelings of desire and attraction and how these are played out in relationships. This form of sexual knowledge was more akin to that found in some women's magazines where tips on 'how to have great sex', 'how to break up with your partner' and 'how to recognise when you are in love/lust' are offered. Such knowledge contrasts with 'official' discourses in its recognition of sexual activity as an embodied, sensual and possibly emotional experience. This discourse contains a sense that sexuality is an integral part of subjectivity and that sexual encounters are potentially desire filled and pleasurable. With these interests at its core this form of sexual knowledge can be characterised as 'a discourse of erotics'.

This discourse surfaced across all of the methods and in the following example emerges in the couple activity when Ngaire responds to a query about the source of her sexual knowledge. Referring to a time before she had engaged in a sexual relationship, she sought information from secondary sources such as experienced friends.

> Ngaire: So when I had my chances at school I'd speak to er my friends that weren't virgins and were virgins. I was really nosey and I was like are you a virgin? And sometimes I'd get a 'yes' and sometimes I'd get a 'no' and uhm *(laugh)* and I'd just say 'oh how'd it feel'?..and you know 'did you tell your mother'? *(laugh)*.
>
> (CA, NAS, 19)

The source Ngaire consults and the questions she asks provide insights into the kind of sexual knowledge which interested her. Her question 'how'd it feel'? suggests a curiosity about the erotics of embodied sensual feeling. Details about how sexual activity might corporeally feel are not usually addressed by sexuality programmes. In fact in an earlier comment Ngaire explained that it was because Health Education had failed to answer her questions that she was forced to seek this knowledge elsewhere.

This conceptualisation of sexual knowledge was also apparent in young people's talk about the sorts of information they wanted sexuality education to convey. In these narratives the distinction between 'official' and 'erotic' discourses was clearly demarcated. It is interesting that although my question schedule had not included sexuality education, young women and men in the couple activity and individual interviews invariably mentioned it when discussing sexual knowledge. Their introduction of this topic within this framework gives support to the fact that they did not discount such information as sexual knowledge. However, many were highly critical of the way that information about embodied feelings, the logistics of bodily interaction and emotional implications was 'missing' from such programmes. In an example of their conceptualisation of sexual knowledge as involving 'official' and 'erotic' discourses Peter and Amy describe how the technicalities of how to have sex were ignored in their sexuality education classes in favour of 'menstruation education'.

> Peter: This is a penis, this is what you do with it...that just never came up [in sex education].
>
> Amy: Yeah a lot of the stuff they cover in sex education is like, this is a pad, this is a tampon, the menstruation stuff...well I mean we covered all that in Intermediate School,[2] in Intermediate School sure, I mean that's a good place to cover menstruation...but once you hit, I mean they are still talking about it at 16 and by that stage you've sort of had a few [periods] and...you don't really want to know about that sort of thing...
>
> (CA, AS, 18)

Sexual knowledge as a discourse of erotics 'this is a penis, this is what you do with it' is clearly juxtaposed here with official discourses of information 'this is a pad, this is a tampon'. The latter being judged as irrelevant at senior school level.

Another couple described how sexuality education had taught them about official discourses of sexual knowledge in terms of the negative consequences of sexual activity while effectively bypassing 'the actual thing'.

> Tim: But at guys schools they don't teach you that [about sex]. Not at my school anyway.
>
> Emma: We learned a lot about what was going to go wrong rather than the actual thing. Like sometimes we would learn more about that, cause they would focus on the STDs and that and not other stuff.
>
> (CA, AS, mixed)

Emma went on to indicate that by 'the actual thing' she meant '...a lot more focusing on not so much the outcome but more on what it involves and how it feels....' (CA, AS, 17).

   Young people's sense of sexual knowledge as more than the negative consequences of sexual activity was also apparent when they expressed what kind of information they had wanted sexuality education provide. These requests sought knowledge about the 'erotics' of sexual activity such as when Emma explained she wanted to know what sex felt like. In Peter and Amy's case, they felt being told more about how to initiate a sexual relationship and how to have sexual intercourse would have made their first sexual experience together, more pleasurable.

Peter: The stuff you want to know is like how is the best way to go about, maybe obtaining condoms or how to approach a girl and say 'hey do you want to have sex', or even how to do it, like just the basics, like positioning or whatever or something like that I mean...

Amy: What makes it, you know with regards to positioning what makes it easier for both you know sexes...If you've got absolutely no idea, you're just going to stuff around for ages and you know most times it's painful, if you don't know what you are on about....

(CA, AS, mixed)

Another couple Nina and Neil, described how they would have preferred less emphasis on the dangers of sexual activity in their sexuality education lessons. While they felt it was important to detail safer sex practices, they also wanted to know more about the pleasurable aspects of sexual activity.

Neil: They make it out to be always they put all these negatives at you that you know, there's so much chance that you'll get pregnant and then your life will be stuffed up and then you can get STD's and that...I think it's to put you off actually and to shock you out of it. They don't sort of, I'm not saying they should encourage all the younger people to go around and have sex like in the 1970s. But they shouldn't make it out to be such a bad thing.

Nina: ..I reckon that even the school nurse should have them [condoms]....in sex education you should actually get some or something or...and be taught how to use one and how sex is not like necessarily a bad thing but if you don't use them it can end up in problems and that.

(CA, AS, 17)

These extracts reveal young people's conceptualisation of sexual knowledge as comprised of official and erotic discourses. Official discourses take the form of information offered in sexuality education about sexually transmitted infections and the negative outcomes of sexual activity, while a discourse of erotics encompasses the lived experience of embodied sexuality. The critical flavour of young people's comments about sexuality education indicated that they felt it had been dominated by official discourses prioritising medical and clinical concerns, while ignoring the positive and embodied aspects of sexuality. Participants indicated that the knowledge they wanted and would have found useful revolved around a discourse of erotics. This included information about 'what it [sexual activity] involves', 'how it feels', how to approach a potential partner and the details of physical intimacies such as putting a condom on. An overarching feature of this discourse as conceptualised by these young people is a sex-positive approach to their sexuality where 'sex is not like necessarily a bad thing' (Nina).

Prioritising a discourse of erotics

In their call for the incorporation of a discourse of erotics within sexuality education the young people quoted above indicated their prioritisation of this sort of information. Given the average age of sexual initiation in New Zealand is 16–17 it is unsurprising that it is erotic knowledge which participants reveal engages their interest and is most relevant at this point in their lives (Dickson, Paul, Herbison and Silva, 1998). The prioritisation of a discourse of erotics was seen within focus groups where it dominated topics of conversation. By contrast official discourses emerged in only 3 of the 17 focus groups, with one group mentioning HIV/AIDS, another unplanned pregnancy and a final group sexually transmissible infections. These issues were raised in relation to the question, 'What problems do young people face in relationships'? This question's emphasis on negative consequences of sexual activity may have sparked these official discourses. A typical example of the way a discourse of erotics surfaced in discussions appears in this conversation from an all female focus group at school. The following is their explanation of how they 'rated guys' on sexual desirability.

Louisa: So what does the rating consist of?

Lita: Umm of 1 to 10 you know so like personality...but that's once you have known the guy for so long, say like 2 years.

Georgia: If you get a whole load of girls together and you get them to talk about their ex boyfriends they will rate them on things like how romantic they were.

Lita: Looks.

Sandra: Physique.

April: Muscles *(laugh)*.

Georgia: Whether you can have a conversation you know and just things like that.

Lita: Yeah and you'd say things like, you know, 'he was horrible but he had nice abs' *(all laugh)*.

Georgia: He had a really good body...but he was so dumb.

April: What about that butt? *(all laugh)*.

(FG, AS, mixed)

Within this talk lies young women's embodied feelings of attraction and desire for the physical appearances and personalities of the young men in question. Embedded within the act of 'rating guys' is an interest in evaluating what is important in a relationship 'whether you can have a conversation' and/or whether 'he had a good body' as well as how they might be treated, 'he was horrible' or he was 'romantic'. These concerns are very much a part of erotic knowledge and were not simply the preserve of young women. Young men's focus group talk also settled on issues of erotics with one common theme being techniques of attracting and asking a potential partner out. As young men are positioned as the initiators of sexual activity within dominant discourses of (hetero)sexuality this element of erotic knowledge has pertinence for them. The narrative below is typical of these sorts of conversations.

Peter: ....it takes a lot of guts just to go up to somebody and say 'hey do you want to go out'. Because there is always a thought that, it could blow up in your face or whatever...in a group you know you can be with them and then you can say to other people 'do you think she likes me sort of thing?'...

Dean: ....well a few situations that I have been involved in where people get friends to test the water for them, like maybe ring the person up or uhm or just ask questions about them.

(FG, AS, mixed)

A discourse of erotics appears here in young men's concern with how to act on the attraction they feel for someone without having to face public rejection if these feelings are not reciprocated. Dean and Peter engage in a sharing of knowledge about how to manage this situation for best effect. In addition to spending more time discussing issues which might form part of a discourse of erotics, young people's prioritisation of this conceptualisation of knowledge was also signalled by their animation and enthusiasm when conversing about these topics. At such times there was much laughter and gesticulation with talk appearing to flow more easily and greater numbers of participants contributing. This animation was not mirrored in discussions of condoms or sexually transmissible diseases which rarely materialised in conversation.

What young people say they know about sexual knowledge

Despite mentioning and appearing to talk more frequently about erotics many young people's questionnaire responses revealed that they felt they lacked knowledge in this area. As explained in Chapter 2

Table 3.1   What Young People Think They Are Knowledgeable About

Items subjects believe they are knowledgeable about	% young women ($n = 263$)		% young men ($n = 148$)		% total ($n = 411$)	
	K	*No K*	K	*No K*	K	*No K*
How to put on a condom	86	14	93	7	88	12
What turns a partner on	74	26	69	31	72	28
Getting what you want out of a sexual relationship	64	36	53	47	60	40
Sexual positions and techniques	56	44	61	39	58	42
How you can contract HIV/AIDS	93	7	91	9	93	8
What lesbianism is	89	11	84	16	87	13
What homosexuality is	90	10	84	16	88	12
The stages and process of conception	74	26	59	41	69	32
What STDs are and how they are caught	88	12	81	19	86	15
How to avoid unwanted sexual activity	83	17	78	22	81	19
What prostitution is	88	12	82	18	86	14

items in the survey had been specifically designed to tap knowledge of both official and erotic discourses. Following the conventions of official discourses of sexual knowledge, questions were posed about STIs, safer sex practice and legal regulations governing sexual behaviour. These questions were interspersed with items which comprised a discourse of erotics in that they asked whether young people felt knowledgeable about 'sexual positions and techniques', 'what turns a partner on' and 'how to get what you want out of a sexual relationship'. The format of these questions required young people to place a tick against items they felt knowledgeable about.

The majority perceived themselves as most knowledgeable about conventional sexual knowledge items, with approximately 9 out of 10^3 indicating they knew how they could contract HIV/AIDS and, slightly less than 9 out of 10^4 reporting they knew how to put on a condom. As demonstrated earlier this type of sexual knowledge forms the basis of sexuality education and as many of the participants had been exposed to such programmes their knowledge in this area was high. Interestingly a higher percentage of males (93 per cent compared with 85 per cent of females) indicated they were knowledgeable about condom use. Given that young men are the ones who wear condoms they may have more experience of how to put one on. The only other significant gender difference on these standard knowledge statements was that a higher percentage of young women (74 per cent compared to 58 per cent of males) felt they were knowledgeable about 'the stages and process of conception'. Within sexuality education this information is often constructed as having greater relevance for young women because it is within their bodies that sperm and egg meet and the zygote develops. This constitution of conception as a 'women's issue', is also evident in calls for sexuality education to be made more relevant for young men as its current emphasis in these areas does not capture their interests (Forrest, 2000). Young men's lesser knowledge in this area could be a consequence of such information being seen as holding less pertinence for them.

The items young people felt they knew least about were those that could be characterised by a discourse of erotics. Forty two per cent of young people reported they had no knowledge of 'sexual positions and techniques', while 40 per cent of the sample disclosed they perceived themselves as ill-informed about getting what they wanted out of a sexual relationship. This sense of lacking erotic knowledge was also apparent in responses to another survey question which asked young people to complete the sentence, 'The kinds of disagreements likely to arise in your relationship(s) about sexual activity are...' While some of these disagreements could be conceptualised by an official discourse of sexuality such as contraception, STIs and pregnancy,[5] the predominance of disagreements related to issues like; where and how often to have sex[6] as well as positions and types of sexual activity.[7] Some examples of survey answers were:

> What positions to have sex in (Q, AS, 16, Male)
>
> Who goes on top (Q, AS, 28, Female)
>
> Wanting to experience different positions (Q, NAS, 17, Male)
>
> When to have it, when not to have it and whether both of us feel comfortable with it. (Q, NAS, 19, Male)

Each of these answers implies that young people find negotiating the logistics of corporeal activity a point of contention and that information concerning how this might be successfully managed may assist them. Responses to both questions echo young people's interest in and desire for knowledge about how to initiate and engage in sexual activity expressed in focus groups and interviews. This desire for 'erotic knowledge' is echoed by young people in another study where questions placed in a 'suggestion box activity' invited the greatest numbers of enquiries about sexual acts. Almost all were posted by young men and solicited descriptions of oral sex, masturbation, anal sex, information about positions for (hetero)sexual intercourse and what were the most pleasurable sexual activities (Forest, 2000). Such information would be considered 'too risky' in many sexuality programmes on the grounds that it may incite sexual activity. Its incorporation would also acknowledge young people as sexual, a subject position which much sexuality education works hard to diminish.

Sources of sexual knowledge

Data were also collected about preferred sources of sexual information providing further insight into the place of 'official' and erotic discourses within young people's conceptualisation of sexual knowledge. Consistent with other studies 'friends' were rated as the most useful source of information (Wight,[8] 1994; Elliot, 1997). This finding arose in answer to a survey question about where sexual knowledge was accessed. It was also regularly mentioned in focus group and interview discussions that 'friends' provide consultation on sexual matters. The fact that young people named

Table 3.2 Preferred Sources of Sexual Knowledge

Information sources	Never consulted (%)	Not useful (%)	Useful (%)	Very useful (%)
The Internet	87	5	5	3
Romantic novels	52	28	17	2
School sexuality education	4	13	51	32
Educational books about sex	29	17	35	20
Pornographic magazines	58	20	12	10
Parents	29	20	28	23
Other family members	38	20	29	13
Magazines for women	37	17	28	17
Television	8	29	48	16
Friends	4	13	41	42

their friends as the most useful source of information may also indicate something about the type of sexual knowledge they seek.

Friends may appear more accessible and less imposing sources of sexual information because they do not signify the authority of adults. Young people may also view them as more approachable with regards to certain sorts of information, particularly erotic knowledge. This possibility surfaced during the couple interviews when five of the six young women[9] mentioned turning to friends for support and details about sexual activity they perceived unattainable from other sources. Significantly more young women than young men found friends a 'very useful' source[10] subsequently it is data from young women that is explored here. This gender difference and additional ways young men might access erotic knowledge are examined later in this section. As seen above, when sexual health education left Nagire with unanswered questions she turned to her school friends to understand things like what 'sex felt like'. In Amy's case she asked her friends to accompany her to a sexual health clinic to buy condoms and to help her ask the nurse questions she might otherwise forget. When Becky needed advice about her experience of sexual activity with her boyfriend, she consulted a friend who was also in a long-term sexual relationship. In addition, Nina explained how she had asked a sexually experienced friend how to perform oral sex on her boyfriend.

Nina: ..you know you don't actually get told how to have sex and stuff. It's just like you're already supposed to know sort of thing and I think that part was all right cause all I had to do was kind of lie there *(laugh)*. But you know like other things, like I said before going down on him. I had to, I just had to find out first. I'd talk to my friend cause she was like what other people would call a slut, but she isn't. She is a bit more experienced than me so I go to her for consultancy *(laugh)* like on this particular matter...

(II, NAS, 17)

In each of these situations young women looked to friends to provide forms of erotic knowledge which weren't available from official discourses of sexual knowledge. As Nina succinctly put it, from official sources, 'you don't actually get told how to have sex and stuff'. It is likely that friends were chosen over sexually experienced others (such as parents) because 'erotic knowledge' is constituted as taboo for young people. This message is signalled through its absence in sexuality education curricula and the moral imperatives young people receive about not engaging in sexual activity. Asking an adult for this information poses greater risk of having this enquiry judged as inappropriate, while friends who are less likely to tow a moral line offer a safer option.

The fact that 'sexuality education' emerged as the second most useful source of information in questionnaire responses, reinforces the idea that young people's conceptualisation of sexual knowledge is comprised of 'erotic' *and* 'official' discourses. This ranking indicates that despite the often heard cry that sexuality education is a waste of time because as Neil puts it '.... it was a period where you sat and just blooming lazed around and made jokes with friends. Cause you knew everything, it was just pointless' (CA, NAS, 17) it does have relevance for many young people.[11] As the majority of respondents correctly answered conventional measures of sexual knowledge taught in these programmes such recall indicates some interest and/or investment in these classes. Another reason sexuality education may have scored highly is that schools from which the sample was recruited were committed to delivering quality programmes which were more likely to be relevant and interesting for students. While young people were critical of sexuality education their ranking of it here suggests that far from discounting such programmes as sources of information, their content formed part of young people's conceptualisation of sexual knowledge.

The way friends outrank sexuality education lends further support to the importance young people assign to erotic knowledge within their conceptualisation of sexual knowledge. If friends are named first and it is likely they provide a source of erotic knowledge, then this is another example of the way in which young people give precedence to this type of knowledge. This ranking also conveys their thirst for erotic knowledge and a sense that it is not being provided by other sources named as useful such as sexuality education and parents (ranked third).

When analysed in relation to other findings from the research, the significant gender difference in consultation of 'friends' implies something further about young people's quest for 'erotic' knowledge. Why young men may not find their friends as useful a source of sexual knowledge emerged within focus group discussion. Participants from three groups raised the issue of young women being more likely to talk with friends about sexual matters. Marcel captured the essence of these comments when he said, 'I don't know...I mean I think...girls are probably better...I don't...I'm not saying that guys couldn't be but, they (girls) just seem to be better at being able to talk about intimate stuff together' (AS, 18 years). The articulation of appropriate masculinity requires that young men 'know it already' in matters of sexuality (Kehily, 2001). The male peer group offers a context in which young men must demonstrate competence and appear knowledgeable about sexual issues in order to achieve masculine status. In this environment talk seeking information about sexual intimacies might be interpreted as sexual ignorance and risks an esteemed masculine identity for the inquirer. Soliciting sexual knowledge from male friends is a precarious process requiring deft identity management which may impede the kind and quality of information gleaned. For example, other studies have revealed how boasting and comic story telling, the format for much male sex talk, offers misleading and inaccurate information (Holland, Ramazanoglu, and Sharpe, 1993). Stories about sexual exploits in which sexual knowledge is implied are often regarded with scepticism by other young men who recognise that exaggeration and lying may be at play (Wight, 1994).

As consulting friends risks masculine identity and may result in misinformation it is likely that young men quench their thirst for 'erotic' knowledge elsewhere. A gender analysis of the questionnaire findings reveal that after 'Friends' and 'Sexuality Education', pornographic magazines and television are ranked equally as the next most useful source by young men. There was a highly significant gender difference[12] between young women and men who found pornographic magazines 'very useful' with only 3 per cent of young women reporting this. Almost three quarters of young women had never consulted pornographic magazines, while almost three quarters of young men had.[13] These findings are replicated in other studies where young men frequently (while young women rarely) sited pornography as a source of sex education (Holland, Ramazanoglu, and Sharpe, 1993). Young men may use these magazines as a source of erotic knowledge about sexual activity. Those in Holland, et al. (1993) study indicated using pornographic magazines and videos as a source of information about the female body (Holland et al., 1993). Mainstream pornography as a source of erotic knowledge is problematic in that it can arguably be seen to reproduce oppressive male sexual subjectivities that sustain inequitable sexual politics. However, in the absence of official discourses which positively describe embodied sexual pleasures, mainstream pornography offers an attractive and convenient source of erotic knowledge.

The sources from which young people access information about things sexual give further insight into their conceptualisation of sexual knowledge. Through their ranking of sexuality education as the second most useful source young people indicate that official discourses of information feature prominently in their constitution of sexual knowledge. The fact that it is friends who are perceived as the most useful source however, suggests that the information they supply which is more likely of an 'erotic' nature is what they perceive as most appropriate to their needs. This prioritisation has some important implications for sexuality education programmes and its strategies. It suggests as educators we need to recognise that the information young people perceive as important and which they most want to know about is 'erotic knowledge'. This is embodied sexual knowledge which as Emma said told you about 'the actual thing' that is, 'what it involves and how it feels' in a way which doesn't 'put all these negatives at you' (Neil) and recognises sexual activity as a potentially pleasurable and positive experience. Young people's descriptions of such knowledge as missing from sexuality education, the way their talk constantly levitated towards it and their turning to friends to obtain it, highlight this as a need to be filled. The findings in the next section suggest that incorporating a discourse of erotics within sexuality education might offer a way of closing the perceived knowledge/practice 'gap'. This is because how young people conceptualise their sexual knowledge has important ramifications for their perception of the relationship between knowledge and practice.

The relationship between knowledge and practice: young people's perceptions

In the remaining part of this chapter I focus on how young people understand the relationship between their sexual knowledge and their sexual practice. This question returns us to the problem of the knowledge/practice 'gap' and is an attempt to examine the implications of young people's conceptualisation of sexual knowledge for the translation of information acquired from sexuality education into practice.

A question which directly probed how young people understood the relationship between their sexual knowledge and their sexual practice was included in the questionnaire. Participants were asked to indicate whether or not they felt their level of sexual knowledge affected their relationships or ability to have relationships. This question was based on an assumption inherent within the 'gap' equation that to engage in sexual relationships 'successfully' some level of sexual knowledge is required. An underlying premise here is that sexual practice without sexual knowledge is more likely to lead to negative consequences such as 'unplanned pregnancy' or the contraction of a sexually transmissible infection. Young people's own views diverged from this logic in that 60 per cent reported that their level of sexual knowledge did not affect their relationships or ability to engage in them,[14] with significantly more young men indicating this.[15] This gender difference might be explained by the fact that if young men are constituted within dominant discourses of heterosexuality as sexually knowledgeable, then the possession of sexual knowledge is less likely to be perceived as relevant to them (that is, it is taken for granted). For these young people the possession of sexual knowledge was seen as having minimal impact on their ability to forge or conduct relationships.

Such a result suggests that one of the reasons some young people do not put their knowledge into practice is because they do not perceive knowledge as necessary for practice. This conceptualisation of the relationship between sexual knowledge and practice is quite different to that contained within the idea of a 'gap' phenomenon. The 'gap' equation presumes a logical progression between firstly having knowledge and then acting upon it. By contrast, for these young people

Table 3.3 Does Sexual Knowledge Affect Young People's Relationships?

	Subjects		
	Young women %	*Young men %*	Totals %
No it doesn't affect relationships	43.0	32.4	39.2
( ) = N	(113)	(47)	(160)
Yes it does affect relationships	57.0	67.6	60.8
( ) = N	(150)	(98)	(248)

knowledge does not appear to be a prerequisite to practice, in fact knowledge may not enter the equation at all.

Young people's perception of the relationship between sexual knowledge and practice might be understood with reference to the way in which (hetero)sexual relationships are discursively constituted as 'natural'. As supposedly 'natural' phenomena, specialised knowledge is not required to engage in them because they are seen to evolve in a preordained fashion. This sense of relationships as naturally occurring in an inevitable course was conveyed in young people's descriptions of how they determined when to engage in sexual intercourse. In a manner typical of other participants Emma expresses this notion of relationships as predestined.

> Emma: ..things had just been like progressing along naturally they were like taking their natural course and getting more and more involved and it just seemed like that was going to be the next step up from what it was before like evolution in a way...
>
> (II, AS, 17)

The idea of sexual activity in (hetero)sexual relationships as a 'natural' 'evolution' also surfaced in focus group discussions. In a mixed gender focus group at school, young people's talk was characterised by references to sexual relationships as 'socially expected' and 'biologically determined' emphasising their preset pattern.

> Louisa: So why get involved in sexual relationships then?
> Rosalind: Just kind of expected of you...experimenting or what ever
> Rodney: Feel like you have got to...you might hear what society is saying, this is what you do and stuff.
> Annalise: mmm...find a mate.
> Tracy: But also it's natural as well.
> Annalise: Yeah.
> Roy: It's animal instinct as well....yeah cause the whole idea of boyfriend/girlfriend relationships is pretty much based around one and one sex wanting to be with another or what ever you know, it's just basic human instinct.
>
> (FG, AS, 17)

Young people's use of language such as 'natural course', 'logical progression', 'the next step' and 'animal instinct' conveys a sense of the natural order of heterosexual relationships and sexual activity within them. As Gavey and McPhillips (1999) note 'the language of the natural evokes the kinds of biological explanations which imply something is "universal, pre-social and essential"' (p. 40). This evocation of heterosexual relationships renders knowledge about sexual activity instinctual and therefore makes its formal acquisition unnecessary. The underlying assumption is that sexual knowledge is not necessary for sexual relationships because the body already knows what to do. Presumably the knowledge which the body 'naturally' knows is not that of how to engage in safer sex, but a knowledge of 'erotics'. While how to put a condom on correctly must be learned in contexts like sexuality education, engaging in sexual activity is constituted as something already known. This conceptualisation of the relationship between sexual knowledge and practice once again reveals young people's prioritisation of a discourse of erotics.

This prioritisation of erotic knowledge presents a problem for the thinking which underpins the 'gap' equation. While it is official discourses of sexual knowledge which are considered necessary for sexual practice, young people in this study did not see the same sexual knowledge/practice relationship. Their prioritisation of a discourse of erotics means that their sense of needing sexual knowledge as the 'gap' phenomenon imagines it, is not the same. With its focus on sexual knowledge as official discourses of sex and sexuality the 'gap' equation fails to recognise young people's own conceptualisation of sexual knowledge as different. If the kind of sexual knowledge which young people indicate they need and seek out is 'erotic' and this knowledge is absent from the 'gap' equation, then the relationship between knowledge and practice is not a 'gap', but rather no relationship exists. This offers a possible explanation for why the majority of participants responded negatively to the survey question about their level of sexual knowledge influencing their relationship practice.

Another way of thinking about why young people in the study appeared not to perceive the relationship between knowledge and practice as a 'gap' is that for them practice *is* knowledge. This returns us to the finding discussed earlier about how participants conceptualised sexual knowledge as comprising information acquired second hand from sources (such as friends and/or books) and that obtained from personal experience. In this way practice acted not only as a means to knowledge but in these young people's minds a form of knowledge, considered to have more status than that gleaned from other sources. This conceptualisation defies the logic of the 'gap' equation which implies sexual knowledge (in the form of official discourses) is necessary for practice, by asserting that it is practice which is most important because it *enables* knowledge. Such findings have important implications for the way in which we might formulate sexuality education strategies and policy.

So what does this mean for sexuality education?

The finding that young people in this study conceptualise sexual knowledge differently from the way it is constituted within the 'gap' equation suggests that as sexuality educators we need to build acknowledgement of this discrepancy into our strategies and programmes. A primary issue here is that within the 'gap' equation sexual knowledge is constituted by official discourses of information while young people's own conceptualisation comprises both official and erotic discourses. I would argue that sexuality education that fails to incorporate and positively acknowledge a discourse of erotics does not take account of young people's own view of sexual knowledge. Participant's critique that erotic knowledge is missing from sexuality education, their sense of lacking such information and frequent attempts to unearth it as well as the way such discourses were prioritised in their talk, indicate their significant investment in this knowledge. Programmes which do not acknowledge how young people themselves conceptualise sexual knowledge and which fail to address their needs and interests as articulated by them, will not succeed in engaging their recipients, whatever their messages.

Young people's conceptualisation of sexual knowledge may also have implications for our understanding of the 'gap' equation and the formulation of sexuality education practices and strategies. Recognising that sexuality education (as a form of secondary knowledge) is not as highly esteemed as knowledge gained first hand, provides some insight into why its messages are not (consistently) translated into practice. If sexuality education is to be successful it must take into account the greater status young people attach to sexual knowledge gleaned from practice. Such kudos is not only associated with the kind of esteemed sexual identity engaging in practice offers, but also the kind of (erotic) knowledge it affords. One way of raising sexuality education's status in young people's eyes as a more useful source of sexual information may be for it to include a discourse of erotics. This possibility is explored in detail in Chapter 7 and would mean opening discursive spaces within programmes for sexual activity to be constituted as a positive, embodied and sensual experience.

What also has implications for sexuality education is that participants did not perceive the relationship between sexual knowledge and practice in the same way as it is envisaged in the 'gap' equation. This discrepancy exists partly because their conceptualisation of sexual knowledge and its inclusion of erotics does not equate with the sense of sexual knowledge embedded in the 'gap' equation. Subsequently, young people do not read the logic of the 'gap' equation as intended (where official discourses of sexual knowledge are seen as a prerequisite to practice). For them, practice provides sexual knowledge, which they consider more useful and to which they attribute greater value than information from other sources. The problem for sexuality education lies in addressing the fact that these young people do not perceive the official discourses of sexual knowledge conveyed by these programmes as the whole picture. If sexual knowledge is more than the information sexuality education offers and this additional information is considered more interesting, the logic of the 'gap' equation falls apart. Similarly, if this additional information, that is erotic knowledge can be acquired through sexual experience then the idea of there being a 'gap' between knowledge and practice makes even less sense. These findings reveal the redundancy of the concept of a knowledge/practice 'gap' in thinking about the effectiveness of sexuality education programmes. They also expose the adult centred constitution of this problem and its disregard for what young people conceptualise as sexual knowledge. Providing young people with 'the whole picture' by enabling a discourse of erotics within sexuality education might enable a realignment of programmes with their own interests as articulated by them.

This chapter has hinted at the relationship between sexual knowledge and young people's sexual subjectivities in the way that possessing or lacking sexual knowledge has implications for revered sexual identities. In the next chapter the way in which young people understand themselves as sexual is explored in more depth in order to elucidate the place of subjectivity in making sense of the 'gap' phenomenon.

4
Sexual Subjects: Young People's Sexual Subjectivities

To view young people as sexual subjects who might legitimately seek sexual pleasure and express sexual desire is controversial. Such contention stems from the fact that this idea conflicts with notions of 'good' youth as sexually innocent. Young people know only too well the negative consequences which threaten if they are identified as sexually active by adult others. These can range from unwelcome warnings about the dangers of sexual activity to closer surveillance and regulation of their activities. Within schooling contexts the pairing of 'student' and 'sexually active' often translates as a 'taken-for-granted signifier of resistance to education' and an indicator that a student will not succeed academically (Nash, 2001). For young women there are particular ramifications when 'femininity' and 'sexually active' are linked. Tolman (2002) explores these in her work around young women's experiences of sexual desire where being seen as sexually active makes it difficult to develop an identity as a '"good", acceptable, moral and normal woman' (p. 44). Such constructions of youthful sexuality vilify and pathologise young people invoking them as 'problems' to be solved (Aggleton and Campbell, 2000, p. 286). There is no recognition within these conceptualisations that young people's sexuality is a positive force which might engender empowerment and provide legitimate pleasures.

This chapter is launched from a different premise: Young people are sexual subjects who have a right to act upon their sexual desires and express and explore their sexuality in positive ways (that is in ways which do not involve the coercion, exploitation or abuse of others). Being sexual is an integral part of being human and a potential source of emotional and corporeal pleasure. Failing to recognise this is a contravention of young people's basic human rights (Pan American Health

Organisation and World Health Organisation, 2000). In line with the stance that young people are meaning givers and construers of their own realities I look here to young people's own articulations of their sexuality and its expression. Examining young people's own sense of themselves as sexual is useful for thinking through the 'gap' equation in several ways. It acknowledges that young people have an active role to play in how knowledge gained from sexuality education is put into practice. This examination also recognises that how young people view their sexuality may have some bearing on what knowledge is acted upon and why.

The social meanings young people ascribe to their sexuality are drawn from dominant discourses of heterosexuality in ways which suggest engagement with, rather than a simple adoption of such meanings. This complexity in the discursive formation of sexual subjectivity is captured by young people's simultaneous accommodation and resistance of dominant discourses of heterosexuality. I propose that how young people talk about sexuality and their presentation of sexual identity during the research process, is contingent upon their access to particular discursive resources and the context in which such articulations occur.

Instead of presuming that young people have a sense of themselves as sexual (something which a plethora of moral discourses work to discourage) the chapter commences by establishing this recognition of sexual self. It then moves to an analysis of the ways in which young people talk about sexual subjectivity by drawing on and resisting dominant discourses of heterosexuality. This discussion exposes the gendered nature of sexual subjectivities and the subsequent possibilities and constraints for sexual expression this entails. The final section provides an account of the presentation of sexual self in the research context by analysing the ways in which young people manage their own sexual identities through their talk about sexuality. Understanding how young people signify their sexual identities within a research situation may offer insights about how sexualities are constituted elsewhere.

Recognising the sexual self

Suggesting that young people may find it difficult to conceptualise themselves as sexual subjects might be conceived as being naïve and ignorant of contemporary youth culture. The realisation that young people are sexual beings underpins the philosophy of much sex education which aims to regulate sexual expression. I propose that this sense of young people as sexual is not their own and instead the effect of (adult) discourses which constitute adolescent sexuality as requiring restraint. The kind of sexual subject produced by such discourses is not an embodied sexual self who knows their own desires and is capable of acting responsibly on them. It is someone who overwhelmed by hormonal urges is incapable of 'rational' sexual decision making and is out of control. For young people who do not recognise themselves within such discursive constructions (or who do not wish to) taking up a subject position *as sexual* is problematic. Such discourses offer young people only minimal agency in determining their sexual lives by providing narrow options for being a sexual subject. That is, a sexual subject who will 'choose' to avoid, delay or modify their behaviour through the use of contraceptives and condoms.

This is not the only reason that young people might find conceptualising themselves as sexual a challenge. Thinking about oneself as a sexual subject can be an uncomfortable experience as it forces an acknowledgement of feelings and knowledge often associated with the private, inappropriate, dirty and/or deviant (Tolman, 1994). These feelings are the result of the way in which sexuality is socially conceived and experienced (Holland, Ramazanoglu, Sharpe, and Thomson, 1998). From a Foucauldian perspective sexuality is not a biological or innate personal characteristic but a product of historically located discourses. Discourses are 'ways of constituting knowledge, together with the social practices, forms of subjectivity and power relations which inhere in such knowledges and the relations between them' (Weedon, 1987, p. 108). Sexuality as an object of knowledge is constructed through multiple discourses which cohere or contradict as they constitute meaning. This contestation occurs because they are intimately connected with manifestations of power which render some 'truer' than others. Those which enjoy privileged circulation are institutionally entrenched and proffer meanings which appear natural or common sense (Weedon, 1987). One such discourse about sexual activity is that it is most satisfying in marriage (or a stable relationship) with someone of the opposite gender. As it is materially based within various religious doctrines this discourse is pervasive and capable of evoking feelings of guilt, deviancy and a need for secrecy around sexual experiences which sit outside of this. Given the preponderance of these sorts of discourses which constitute sexual activity as illicit, personal and embarrassing, it is unsurprising that young people may find thinking about themselves as sexual difficult. This difficulty is glimpsed in other research which notes some young people's struggle to articulate their thoughts and feel comfortable speaking about sexual issues (Wight, 1994; Holland et al., 1998).

Discourses which constitute youth as either 'sexually innocent' or any expression of their sexuality in terms of 'deviancy' and 'promiscuity' deny them a positive discursive space from which to be legitimately sexual. Acknowledging oneself as concurrently 'young' and 'sexual' necessitates a reconstitution of these subject positionings and invites young people to conceptualise themselves differently. Given that not all young people had access to alternative ways of describing their sense of sexual self they sometimes found it a challenge to respond to my questioning about being a sexual person. This was apparent in the following focus groups where feedback was sought about the experience of answering the questionnaire.

> Louisa: How did you find filling out the questionnaire?
> Kiri: That was uhm, I'd never thought about half those questions before.
> Alex: Some of the questions that you come across like what do you think of your body, it's like skinny, fat, overweight, sexy, it's like uhm....? [As if thinking...*laugh*]
> Jackie: Like rate your sexiness *(laugh)* Uhm I dunno *(all laugh)*.
> Louisa: So how do those questions sort of make you feel?
> Alex: Some of them made you have to think.
> Nadine: Yeah, yeah like and you just don't know what to answer. (FG, NAS, mixed).

Part of young people's dilemma here may have stemmed from the fact that these questions positioned them as sexual subjects in a way which presumed this was normal and not immoral. Questions which asked about their feelings of 'sexiness' and personal experience of sexual desire worked to give acknowledgement of a positive sexual subject. Alex's claim that she had to 'think' (and by implication hard) may symbolise the rarity with which young people are given the opportunity by adults to think of themselves in this way. Similarly, Kiri's protestation that she'd 'never thought about half those questions' signalled her unfamiliarity with being acknowledged positively as a sexual person.

This sense of it being unusual to treat young people as legitimate sexual subjects was reinforced during a conversation with a teacher in the field work phase of the research. Prior to consenting to the research

Mr Red had requested a copy of the questionnaire and had called me into his office to express his concerns around questions asking participants to describe their sexual desire and how sexy they thought they were. Mr Red wanted to know why such questions were necessary and what possible use this data could yield. Evoking young people as 'sexy' and 'desiring' subjects in a setting whose official culture seeks to deny the sexual and desexualise relations appeared 'dangerous' to Mr Red (Epstein and Johnson, 1998). Not only might it provoke complaints from parents but it could also incite public controversy. It also threatened the disruption of what Fine has described as the 'missing discourse of desire' within schooling contexts by assuming and acknowledging young women and men's embodied sexual experience (Fine, 1988). The symbolic boundaries shaping sexuality in school mean that any acknowledgement of desire is confined to warnings about curtailing its expression or protecting against unsolicited forms, such as sexual harassment. In this environment there is no room for acknowledgement of desire as a positive and pleasurable force in young people's lives and subsequently a denial of students as legitimate sexual subjects. When positioned differently by a questionnaire which assumed their positive status as sexual subjects some young people understandably struggled to register themselves within such a construction.

In this chapter I want to suggest that to transcend the dilemmas presented by the 'gap' phenomenon it is imperative that young people are recognised as legitimate sexual subjects. By this I mean subjects whose sexuality is not perceived as requiring restraint or necessitating risk management. As legitimate youthful sexual subjects young people's sexuality might be viewed as a positive resource from which they could derive a rich source of corporeal/emotional pleasure and satisfaction. It is in offering young people spaces in which to legitimately and positively seek/experience sexual pleasure and desire that responsible, empowered and satisfied sexual subjects can emerge.

Describing the sexual self

Although this was not always an easy project, young people did reflect upon their sexual subjectivities during the qualitative methods and through their questionnaire answers. High response rates to questions pertaining to subjectivity in the questionnaire (such as rating how sexy they thought they were and describing their sexual desire of which 97 per cent and 98 per cent of participants answered respectively) also demonstrated this ability to conceptualise themselves as sexual. Most subjects described themselves as 'sort of sexy'[1] rather than 'very sexy' and a majority reported they had 'average desires'[2] as opposed to 'very strong'. These responses sit uneasily with discourses where adolescence is constituted as a time of raging hormones and surging desires and instead suggested a more sedate sexual subject (Aggleton and Campbell, 2000). This was also reflected in participants' selection of words to describe themselves as sexual partners with the most frequent descriptors being 'fun loving' followed by 'caring' and 'romantic'.[3] In contrast, few opted for words that communicated a more overtly sexual subjectivity like 'raunchy', 'kinky' or 'lustful'.[4] These adjectives characterised young people as sensitive and thoughtful partners who were interested in more than just sexual adventure. Such findings imply that images of youthful promiscuity and sexual deviance do not always capture young people's own construction of sexual self.

When young people talked about their sexual selves during the individual and couple interviews this articulation was inextricably tied to being 'sexually active'. Such enunciations of sexual subjectivity emerged most often in discussions about engaging in sexual activity with someone else, suggesting that what defined young people as sexual was the sexual relationships and encounters they experienced. This association between being sexually active and being a sexual person appears in the following extracts where young people in relationships described how they perceived themselves as sexual people during the individual interview.

Here Ashby, Becky, and Amy describe how their sense of themselves as sexual people was dependent upon whether or not their partners found sexual activity with them pleasurable.

Ashby: ..I'd say it varies like sometimes I think you know, oh yeah she had a good time and other times I think uhm you know I'm not really pleasing her [His girlfriend]...

(II, AS, 17)

Amy: ...sometimes I sort of think oh that was, I didn't particularly perform very well there...I mean Peter will usually say afterwards 'oh you know that was okay, that was good, I enjoyed that', something like that which makes me feel okay, he did enjoy that, it's okay you know he's given me pleasure, okay that's cool I've given him something too....

(II, AS, 18)

Becky: ...I don't really know how he [her boyfriend] views it but I sort of have to think that it can't be bad for him otherwise

he wouldn't be with me and he wouldn't have come back
to me twice you know what I mean. If it was bad for him
then I am sure that he wouldn't be there.

(II, AS, 17)

Throughout the qualitative methods there was a paucity of talk that
indicated being a sexual person was distinct from being a sexual
partner. Narratives which provided insight into sexual subjectivity
involved an omnipresent 'other' against whom a sense of sexual self
was formed. In the narratives above the sexual self emerges in relation
to the pleasure given to a partner during sexual activity and takes on
an aspect of 'performance'. Below the importance of the other partner
in fashioning this sense of sexual self is apparent when Cam and Nina
reveal how different relationships influenced their perception of
themselves as confident/inexperienced.

Cam: I feel confident in mine and Chris's relationship with everything. I think I feel more confident because he isn't, and
I think if we were fairly insecure (*laugh*) it would be really
terrible. But uhm I think that I am more confident with
him than I have been with other people.

Louisa: So you have a positive feeling about yourself as a sexual
partner?

Cam: Yeah, yeah.

(II, NAS, 19)

Louisa: So how do you see yourself as a sexual partner then?

Nina: Well I think that ah, we have both kind of developed
together...I think we are both good for each other kind of
thing. Like if I had another, say if I broke up with this, with
Neil and went into another relationship I wouldn't feel the
same.

Louisa: Different how?

Nina: Well first of all I wouldn't go into another relationship and
start having sex straight away sort of thing and then even
when I did I would probably feel kind of like I did with
Neil inexperienced sort of even though, even though you
know....

(II, NAS, 17)

The idea that sexual practice with a partner provided the reference
point for conceptualising oneself as a sexual person, ties into young

people's hierarchalisation of sexual knowledge described in Chapter 3. With knowledge derived from practical experience held in highest esteem, this knowledge has the power to bestow the status of 'sexual person'. Without practical sexual knowledge and only an intellectual grasp of sexual activity, young people in sexual relationships were less likely to conceptualise themselves as being 'truly' sexual. These narratives also suggest that being a sexual person was perceived to occur at the point of becoming sexually active with someone else and not a core dimension of humanity that is perpetually present. Referring only to oneself as sexual in conjunction with talk about sexual relationships or sexual activity with others implies these young people may not have a sense of sexuality separate from this.

This finding may have significant ramifications for young people's sexual subjectivities given gendered relations of power. Reliance on the occurrence of sexual activity or a partner's perceptions for a sense of sexual self could have gendered constraints for the experience of being a sexual subject. This can be explained with reference to Holland et al.'s (1998) concept of 'the-male-in-the-head' which encompasses a notion of the inextricable relation between sexual self and other. Within this concept 'the other' is found in the form of a male gaze which is played out in a myriad of ways through various layers/levels of power and regulates the sexual expectations, meanings and practices of both men and women. As a result of this operation of power, heterosexuality becomes 'thinly disguised masculinity' so that being a sexual person is confined by the privileging of masculine meanings and desires. For young women in the current study, this means in a very literal sense through the material presence of the words or deeds of a male partner as indicated in the narratives above. At a more intangible level this occurs through the obligatory presence of 'the-male-in-the-head' which Holland et al. (1998) maintain is an integral feature of heterosexual relationships. In this way young women's understanding of themselves as sexual can be likened to a 'woman with a man inside watching a woman' (p. 19). The operation of 'the-male-in-the-head' also maintains unequal sexual relations of power in the way young women must collude in the perpetuation of male power in order to appear appropriately feminine.

For young men the effect of the 'male-in-the-head' is different. On the one hand it cements a privileged position in gender/sexual relations because of the way in which acquiring successful masculinity necessitates an exercise of power over women (and subordinated masculinities) (Herek, 1987; Connell, 1995). Conversely, it produces pressures to conform to hegemonic masculinity or culturally upheld ways of being male which structure relations of dominance and subordination between groups of men (Connell, 1995; Mac an Ghaill, 1996). In these interactions heterosexuality assumes a dominant status while homosexuality assumes a subordinate position. This sex-gender hierarchy encourages young men to adopt cultural ideals of Western masculinity in which men are produced as active, virile, competitive and violent. As Holland et al. (1998) poignantly write;

> Heterosexual young men embark on sexual activities with women in social situations in which they are under pressure to become victorious gladiators in the sexual arena, while avoiding the many pitfalls that can reduce them to the ignomity of being a wimp, a failed man, a sexual flop. (p. 150)

Failure to achieve the culturally defined requirements of hegemonic masculinity exposes young men to considerable vulnerability and may result in them experiencing a number of fears and uncertainties about their sexual selves. Ashby alludes to one of these above when his description of establishing whether his girlfriend found sex pleasurable signals his evaluation of his sexual performance. The demonstration of sexual competence in the form of knowing how to please a woman is a familiar marker of successful masculinity and carries the power to incite performance anxieties. Sustaining an image as sexually 'knowable' can be an enormous weight for young men especially when studies suggest they are less knowledgeable about sexual matters than young women (see Chapter 3). If heterosexual encounters are structured around the presumption of an active and knowing male partner (of which in reality there is an absence) this has potentially serious implications for the practice of safer sex. Another corollary of the 'male-in-the-head' is that it can confine young men's expression of sexuality to those forms sanctioned by the operation of hegemonic masculinity. To step outside these is to risk ridicule from one's peer group and become disempowered within the conventions of masculinity. In this way the male-in-the-head limits the possibilities for how young men might express and experience their sexuality and the pleasures procured from this.

In the ensuing sections I explore in greater depth the power of the operation of hegemonic masculinity in producing young men as sexual subjects. This forms part of an examination of the ways in which young people engage with traditional discourses of heterosexuality in their talk about themselves as sexual. In order to elucidate the theoretical underpinnings of this discussion, I first outline the relationship between subjectivity, power and discourse.

Discourses and sexual subject positions

Given that it is young people's own understandings of themselves as sexual that are prioritised in this book I look here at the social meanings they ascribe to their sexuality. I am particularly interested in the ways in which they become constituted as sexual within talk and the discursive resources they draw upon.

From a Foucauldian perspective contemporary (hetero)sexualities are the product of particular discourses which are articulated around a cluster of power relations (Foucault, 1976). Therefore there are competing possibilities for giving meaning to the sexual self with some having greater currency than others. This occurs because of the way discourses can position subjects by making available particular 'subject-positions' for them to occupy. These subject positions (or certain ways-of-seeing the world and certain ways-of-being in the world) constrain or facilitate sexual thought and practices (Willig, 2001). Willig (1999) explains this power of discourse with reference to Hollway's idea of the 'male sexual drive' in which male sexuality is prescribed as biological, need-driven and asocial. The subject position available to men within this discourse is that of an 'instinct-driven', aggressive, 'sexual predator' (Willig, 1999, p. 114). While this does not preclude men from thinking of themselves in other ways (for example being courted or seduced) it requires their access to other discursive constructions of male (hetero)-sexualities.

The process of 'taking up' subject positions is not simply, a cognitive choice, but rather a complex process of becoming that involves being subjected *to*, and subject *of* discourse (Jones, 1997, p. 261). Butler (1990) explains it this way: 'There is no self...who maintains integrity prior to its entrance into this conflicted cultural field. There is only the taking up of tools where they lie, where the "very taking up" is enabled by the tool lying there' (p. 145). For some theorists however, 'the mere availability of subject positions in discourse cannot account for the emotional investments individuals make in particular discursive positions' (Willig, 2001, p. 118). Some of these theorists employ psychoanalytic concepts to explain subjective affiliations or why individuals take up some 'tools' and not others (see Hollway, 1989).

Others have pointed to the way in which discourses are strongly implicated in the exercise of power and carry different social significance deeming some more pervasive than others. Dominant discourses such as the male sexual drive have greater social power because 'they have firm institutional bases and are so ubiquitous that they appear to be natural, universal, inevitable: simply "common sense"' (Potts, 2002, p. 16). By contrast discourses of the 'new man' have a narrower circulation and do not enjoy the same kind of institutional entrenchment. As dominant discourses legitimate existing power relations and structures by defining what is 'normal', alternative or 'oppositional' subject positions are not usually perceived as desirable or even possible alternatives (Davies, 1989).

There is a raft of dominant discourses of (hetero)sexuality which impinge upon young people's sense of sexual subjectivity. These have been historically shaped by fields such as religion, medicine, law, media and academic disciplines (Hawkes, 1996) and subsequently have strong institutional bases. (Hetero)sexuality as it is invoked through gender arrangements is premised upon 'an essentialist biological/reproductive discourse of male and female sexuality' (Potts, 2002, p. 43). This renders men as 'the knowing sexual agent and actor' in (hetero)sexual relations while women are cast as 'unknowing and acted upon' (Holland, Ramazanoglu, Sharpe, and Thomson, 1994, p. 127). Such essentialist discourses constitute men's sexual desire as insatiable and unbridled while they are seen as possessing bodies which are easily gratified. Female sexuality is constructed in terms of more muted sexual desires and a body which requires additional effort to achieve sexual pleasure. Research which has sought to understand (hetero)sexualities at the level of the micro politics of sexual relations indicates that these discourses offer young men access to an empowering sexuality that privileges their desires and needs (Holland et al., 1998). By contrast female sexuality takes on a subordinate status and is experienced through '.....a protective discourse which emphasises reproductive capacity and the dangers of sexual activity' (Holland et al., 1994, p. 127).

While young people's talk may contain understandings of their subjectivity which are informed by these dominant discourses of (hetero)-sexuality I argue that their narratives are not determined by them. The power of dominant discourses is not monolithic and is constantly challenged by the presence of alternative meanings of (hetero)sexuality. In Foucault's (1980) words '....as soon as there's a relation of power there's a possibility of resistance' (p. 13). With regard to the discursive production of young people's sexual subjectivities this resistance is not the result of a pre-discursive humanist subject who is able to choose their sense of sexual self. Instead it lies within the constitutive force of discourse (Davies, 1997) which is never completely realised and therefore its power never absolute (this would imply *force* not power and subsequently the total determination of the subject). The constitutive force of discourse produces an (inherent) agency for the subject rendering the potential for resistance ever present. As Hekman (1995) explains, 'agency can be seen as a capacity that flows from discursive formations' (p. 203).

One of the aims of this section is to explore Foucault's notion of the possibility of resistance in relation to the discursive construction of young people's sexual subjectivities. I am interested in determining whether young people's talk about their sexual selves resists some of the dominant meanings about (hetero)sexuality referred to above. In this way I attempt to grapple with questions within post-structural theory around the agency of the subject and the power of discourse.

Dominant discourses of (Hetero)sexuality: Young women

As discursive practices that support an active male and passive female sexuality are deeply embedded within social and political participation they regularly featured in young people's talk about their sexual selves. In drawing on these discourses young women were positioned as sexually vulnerable and less easily pleasured than young men, victim to male sexual gratification and more interested in the emotional aspects of physical intimacy. They subsequently appeared as the subordinate partner in (hetero)sexual relationships who were 'acted upon', rather than 'acting'. These conventional conceptualisations of female sexuality are evident in the following narratives from focus groups and individual interviews. In the first example one young woman in a mixed-gender group draws on conventional meanings of female sexuality by stressing the importance of emotional intimacy and her apathy with regard to sexual activity:

> Caitlin: ....but to me it's really the emotional side which is important and that's why I like to cuddle and that rather than have sex...like sometimes I just can't be bothered and just want to get it over and done with....I guess that sounds quite odd but like we have been together for 3 years now.
>
> (FG, AS, 18)

In another mixed-gender focus group where participants were learning business skills, Alex spoke about female sexuality in terms of traditional notions of vulnerability where women's romantic ideas of love made them susceptible to exploitation by male partners:

> Alex: I think like that girls get more vulnerable like being.....
> Leanne: They fall into more....
> Alex: Yeah being in love with the fact of being in love sort of thing and they'll do anything for their man, and the man sort of goes 'oh yeah, take advantage of this'. (FG, NAS, 17)

During an individual interview Cam explained how she saw women as less easily sexually aroused and more likely to be stimulated by foreplay than sexual intercourse.

> Louisa: Is there anything about the sexual part of the relationship that you would change if you could?
> Cam: It's like uhm, just like, guys don't really need as much foreplay as women do to feel satisfied and like sometimes Chris's really good with stuff like that both other times he's not and....
>
> (II, NAS, 19)

When talking about their decision to have sexual intercourse with their partner for the first time, four of the six young women who participated in the couple interviews constructed notions of traditional female sexual passivity. This was seen through their expression of anxiety about having sex and the fact that it was their partners who broached this subject first. Such talk suggested that they did not perceive themselves to necessarily be in control of events. Amy and Emma indicated their anxiety about engaging in sexual intercourse in terms of worries about pregnancy, what others might think, and feelings of insecurity within the relationship. These anxieties implied a reluctance to engage in such activity, even though both young women eventually did:

> Louisa: So how did you feel about having sex?
> Amy: (A) I was worried about my parents finding out. (B) I was worried that it was going to hurt and (C) I was worried about pregnancy and it was really worrying me and it was like well you know I don't know if I really want to do this.....
>
> (CA, AS, 17)

Emma: I'm a bit more wary about, okay very wary about being used and I was still worried okay is he using me then leaving me. Cause I am very dependent on feeling loved and if I was going to enter into that sort of level of relationship [sexual intercourse] ....I was thinking right I don't want to be used again, I want to really know that he loves me....

(II, AS, 17)

This talk draws on conventional discourses of (hetero)sexuality in which women are positioned as the (reluctant) recipients of male desires rather than the initiators of sexual activity. In taking up subject positions offered by these discourses young women are constituted as the objects of sexual attention who must be reassured/convinced that intercourse will not have negative repercussions for them. In the first extracts, Cam, Leanne, Alex, and Caitlin's talk reproduces traditional notions of female sexuality in the way they are seen as less interested in/or easily satisfied by physical contact than young men and instead more concerned with the emotional aspect of relationships.

Dominant discourses of (hetero)sexuality: young men

One of the consequences of drawing on dominant discourses of (hetero)-sexuality in the constitution of sexual self is that it results in gender polarised identities. For young men taking up subject positions offered by dominant discourses meant being seen as sexually assertive, emotionally detached, with a voracious sexual desire and a body that guaranteed them satisfaction. This constitution of sexual subjectivity is a result of the way masculinities are inherently relational and structured around an assertion of difference from femininity (and subordinate masculinities) (Mac an Ghaill, 1994, Connell, 1989; Nayak and Kehily, 1996). Their consequent manifestation as opposite to female sexuality is seen in the examples of talk which follow.

One of the many instances in which young men portrayed an image of traditional male sexuality was when I asked Ashby, how he knew he wanted to have sex with his girlfriend and he explained: 'Uhm I was horny [laugh]. That was about it [laugh]' and later during the individual interview he added '...cause see like I didn't care, I could have done it at the start of the relationship, I didn't need time. I'm a guy' (AS, II, 17). This statement stands in stark contrast to the lack of certainty, sense of anxiety and absence of feelings of strong desire which permeated this decision for the young women above. In their research, Holland, Ramazanoglu, Sharpe, and Thomson (1996) attribute this gendering of the experience of sexual intercourse to the way that for a young man 'this is an empowering moment of symbolic and physical importance, whereby through a physical performance, his identity as a man, and, therefore, a competent sexual actor is confirmed' (p. 158). The situation for young women is somewhat different. Engaging in sexual intercourse can 'break' rather than 'make' an appropriately feminine reputation as well as increase feelings of disempowerment caused by fears of 'being used' or getting pregnant.

Most of the examples of young men taking up these subject positions were found in focus groups or individual interviews rather than in front of their female partners who may have curbed this display of 'hard masculinity'.[5] The first three extracts from young men conceptualise male sexuality as perpetually ready for sex, virile, and potent:

Michael: Guys are basically always ready. (FG, NAS, 19)

Anabella: I heard some statistics.... and guys supposedly think of sex, six times an hour on average.
Darren: Oh it's heaps more than that [all laugh]. (FG, AS, 17)

Tina: I have some female friends who are pretty ready to go [have sex].
Barnaby: [Cheekily] Can you introduce me? [all laugh]. (FG, NAS, 18)

Young men's emotional detachment was communicated by the following young man who implied that the best thing about relationships was their financial benefits.

Louisa: What are some of the best things about being in a relationship?
Vete: Source of income [all laugh]. (FG, NAS, 19)

Tim and Chris who participated in the individual interview were positioned as traditionally masculine through the constitution of their bodies as 'pleasure machines':

Tim: ....if I wanted to ejaculate I could probably just do so in less than a minute. (II, NAS, 18)

Chris: .....a guy is sort of almost guaranteed to feel good [having sex] you know, feel the same in the end anyway so. (II, NAS, 19)

These narratives can been understood with reference to the idea that 'differentiated heterosexual masculinities are produced and inhabited through the collective actions of boys as they "handle" or "negotiate" their concrete social environment, and through relations of similarity with and opposition to other groups within that social environment' (Mac an Ghaill cited in Redman, 1996, p. 174). To achieve full masculine status young men must separate themselves from homosexual and feminine identities. In the above quotations, young men took up subject positions in opposition to female sexuality in their assertion of themselves as all those things female sexuality is traditionally not, 'virile' (Barnaby, Michael), 'sex obsessed' (Darren), easily physically pleasured (Chris, Tim) and emotionally detached (Vete). In so doing they contribute to the perpetuation of hegemonic masculinity which Hearn and Morgan (1995) describe as '… the dominance within society of certain forms and practices of masculinity which are historically conditioned and open to change and challenge' (p. 179). Hegemonic masculinity is not so much a 'type of masculinity', but a form of power which sustains gendered inequality because of the way it achieves the consent of a majority of men who support it. Some of the young men in this research recognised their own collusion in the (re)production of this form of power as indicated in the following narrative:

> Darren: Guys have got a lot to prove. There's a lot … there's a lot for guys to live up to like uhm gotta be all macho and gotta be cool and all this sort of stuff, gotta score nice chicks or if you have got one chick, you have got to score often. (FG, AS, 18)

This recognition of 'appropriate' forms of masculinity did not mean that young men always did, or could, oppose hegemonic masculinity's operation. However, as the ensuing sections indicate, some young men were more likely than others to take up subject positions which did not conform to expressions of masculinity promoted by this hegemony.

Young women resisting dominant discourses of (hetero)sexuality

At any historical moment discursive formations are multiple and heterogeneous, so that 'even though in every era there will be hegemonic discourses, other non-hegemonic discourses will also exist, forming a discursive mix from which subjectivity can be constructed' (Hekman, 1995, p. 203). Contemporary meanings, about for example 'girl power', offer young women the possibility of being positioned as active and desiring sexual agents. As Jones (1993) explains:

> ... the point is that the social order within which femininity is dis-
> cursively constructed ... is not seamlessly consistent ... it is in the
> gaps opened by this unevenness that the possibilities for resistance
> ... can be developed.   (p. 161)

This heterogeneity in discursive formation meant that the sexual subjectivities of young women in this study were not always constituted
through dominant meanings about female sexuality.

A significant minority of young women drew on discourses which
resisted traditional meanings about female sexuality most of the
time, or some of the time in their talk. Rosalind was one of these
young women who had been exposed to other ways of conceptualising female sexuality through attending a school renowned for its
alternative pedagogies. Her access to a set of meanings, which she
named as 'stereotypes', enabled her to resist the positioning of
young women as always wanting commitment and love from relationships. Talking about how sexuality is gendered, Rosalind said,
'I mean you have got your stereotypical, women want commitment
and love and guys just want a fling, but I think that girls are pretty
much like that as well [laugh]' (FG, 17).

Some young women resisted other subject positions offered by the
operation of the sexual double standard. Lees (1993) describes this as
the social processes by which young women who have many sexual
partners are labelled 'slags' or 'sluts', while young men displaying
equivalent behaviour gain the status of 'studs'. Two young women in
the focus groups, Shona and Anna, spoke about the sexual double standard in a way that simultaneously resisted and accommodated the
'slut' label.

> Lindel: My friends every weekend they're with someone different ...
> Shona: I'm one of them ... nah I'm not at the moment though
>   [Shona was about to have a baby in 3 weeks time].
>   (FG, NAS, 17)

Shona's talk illustrates an accommodation of this meaning of female
sexuality in the way she adopts the term 'slut' to describe herself. Her
comment 'I'm not at the moment though' indicates this is not an
understanding she has of herself while pregnant. The inclusion of, 'at
the moment' suggests she might choose to return to being a 'slut' some
time in the future. Her words defy the negative social constitution of
her as a pregnant 17-year-old, by rendering her more legitimately as

'a mother to be'. At the same time her sexual agency is maintained by suggesting that being sexual (that is, sexually active) is something she can return to later. Such talk implies a complex accommodation and resistance of the subject position 'slut'.

Similarly, Anna draws on a discourse of female sexuality which legitimates young women's desire and enables her to resist being positioned as a 'slut', by constituting her sexual desire as 'normal'.

> Anna: I was called a slut when I cheated on someone and I was called a slut ... but a slut is supposed to be someone who sleeps around and I don't sleep around.
>
> (FG, NAS, 17)

By constituting her sexual self through discursive resources that recognise and legitimate women's sexual desire as normal, Anna's behaviour is not defined as that of a 'slut'. Instead her desire for *one* other person outside of her relationship is constructed as ordinary and acceptable. She resists the positioning of 'slut' through talk which constitutes her as a woman who has a right to possess and act on her sexual desires.

A special set of circumstances formed the discursive context within which these resistant constructions of female sexual subjectivity were deployed. Both Shona and Anna were pregnant and living in accommodation for young women adjusting to unplanned pregnancies at the time of the research. In this environment they were given excellent support, counselling and advice about their pregnancy and future life with their babies. As part of this programme they recounted their participation in a workshop where the facilitator had asked them to reflect upon and discuss gendered sexual stereotypes. It is likely their resistance to dominant sets of meanings around female sexuality was made possible by their social location within this environment which enabled access to other conceptualisations of female sexuality.

Dominant discourses of female (hetero)sexuality offer subject positions where women are conceptualised as less desiring and less easily pleasured than men. However, some young women articulated their desire and experiences of sexual pleasure during the research. Their words contested the image of young women as sexually passive, uninterested in sexual contact and unable to enjoy corporeal pleasures. Instead, they described passion and pleasure as normal expressions and experiences of their sexuality. This is revealed in the following discussions within focus groups and couple interviews.

One of the couples Nina and Neil, explained how traditional gender norms which construct males as active and females as passive, were reversed in their relationship:

Neil: Cause like most of the time I can go without it, whereas Nina can't [Neil laughs].
Nina: Yeah it's kind a like, we always hassle that it's kind of like, do you watch *Married with Children?*[6] And I'm Peggy Bundy and he's like Al [laugh] I think that's actually quite a good way to describe it [laugh].

(CA, NAS, mixed)

Unconventional expressions of female sexual pleasure and desire also emerged in focus groups such as this all female group who were at school:

Louisa: ... Yeah because as you say there is that kind of stereotype where it's the guy who always wants the physical activity.
Hine: Yeah but it's sometimes opposite. Heaps sometimes [all laugh].
Louisa: In your experience?
Hine: Yeah.

(FG, AS, 17)

One young woman who was at school and took part in a mixed focus group talked about the importance of the physical side of her relationships in an unconventionally feminine way:

I reckon that like the physical part of the relationship is really important to me like I wouldn't be able to, you know even if I loved someone I wouldn't be able to stay with them for the rest of my life if the physical side wasn't good you know.

(FG, AS, 18)

In another single-sex focus group at school, Lesley openly expressed her feelings of desire and the need to act on them:

Lesley: ... this is just from my experience but, if I feel lust for someone then I ... I have to do something about it.

(FG, AS, 17)

Similarly during a mixed focus group, Rosalind constituted an active female sexuality by explaining how she had decided to ask someone out: 'I was a bit nervous about asking someone out ... but the last time I approached someone it was okay ... It was okay I obviously don't have a problem with it.' (FG, AS, 17) Because of the sexual double standard and young women's need to safeguard their sexual reputations, talk about female desire and pleasure occurred mainly in environments where young women felt they would not be negatively stigmatised. Exclusively female groups where the risk of reputation was equal for all participants, or mixed sex groups where trust between peer members was established, provided examples of 'safe' spaces. Young women's access to alternative ways of constituting female sexuality was also apparent from the quantitative data. When the questionnaire asked what they felt they had the most control over in their sexual relationships (kinds of sexual activity, contraception, how often to have sex), young women reported feeling they had more control than young men over all of the elements of sexual activity named. Of the young women, 78 per cent felt that they had control over contraception, indicating a particular sense of agency around this aspect of sexual activity. These results imply the presence of more active, rather than passive, female sexual subjectivities amongst the women in the questionnaire sample.

Such findings raise important questions about the relationship between how young people's talk constitutes them as sexual subjects and their actual sexual behaviour. Discursive constructions can be seen to have 'real effects' in the way that language constitutes meaning and possibilities for practice. However as Willig (2001) points out: 'If discourse does, indeed, *construct reality*, then to what extent can "reality" be said to constrain discourse?' (p. 119). To illustrate, it might be argued that there is a juncture between the *feeling* of control over contraception in a relationship and actually having access to material power in this situation. This point is supported in Holland et al.'s (1998) research which reported some young women as empowered at an intellectual level but unable to achieve this agency (always or at all) within their relationship practice. Similarly, there is an abundance of research[7] indicating that sexual coercion and rape are experienced by high proportions of women. Such findings suggest that constructions of active female sexuality may not always find expression within the material constraints of relationship practice.

Young men resisting dominant discourses of (hetero)sexuality

While it is important that research draws attention to the oppressive ways masculinities are constructed, it also needs to be attentive to the ways, contexts and times in which boys inhabit alternative (not necessarily subordinate) masculinities and the attractions of these to them. (Frosh, Phoenix, and Pattman, 2002, p. 73)

The aim of this section is to explore the full repertoire of young men's sexual identities including positionings which resist dominant discourses of (hetero)sexuality and young men's investments in these. Conceptualising these as *resistance* to dominant discourses does not necessarily render them a subordinate status. The growing prevalence and acceptance of such performances was highlighted by a young student discussing these narratives when she ventured 'but aren't these alternative expressions of sexuality just normal these days'. The emergence of 'softer' masculinities within other research lends further support to the increased circulation and normalisation of such discursive constructions (Redman, 2001; Walker and Kushner, 1997; Edley and Wetherell, 1997; Thomson, McGrellis, Holland, Henderson, and Sharpe, 2001; Barker, 2000; O'Donnell and Sharpe, 2000). What is transgressive about these sexual subjectivities is the way in which they contest dominant meanings about being male and sexual and subsequently open other possibilities for young men to experience their sexuality.

In opposition to notions that appropriate masculinity involves 'sexual conquest over women and separation from emotional involvement with them' (Holland, Ramazanoglu, and Sharpe, 1993, p. 17) some young men denied sexual intercourse as a primary motive for entering into, or remaining in relationships. In the following discussions young men talk about how sex was not the most important aspect of their relationships.

Richard was in a predominately male focus group (one female) where participants were training to be farmers:

> Louisa: Is the sex an important part of the relationship?
> Richard: Yeah it's part of it aye but it's not you know just what
>   you are there for.
>
> (FG, NAS, 17)

Marcel made this comment while at school in a mixed gender focus group:

> Marcel: It's different for, I mean you watch TV and it's just sex
>   you know, yeah, sex, but it's not just about that really.
>
> (FG, AS, 17)

Darren was also at school and a participant in a mixed gender group:

> Darren: It's [sex] it's not the be all and end all really. (FG, AS, 18)

As a way of emphasising that sex was not of primary importance in their relationships, the following four young men who participated in the couple interview described how they would remain in their current relationships even without sexual activity:

Peter: Yeah, yeah I mean sex is good, it's nice but it's not, it's not essential. I'd still love her, I'd still want to be with her. So you know I mean that it's nice but I mean if it had to stop then it would, and I would still go out with her.

(II, AS, 18)

Chris: I don't actually think that the sex part affects uhm the relationship. If I couldn't have sex with Cam, well I would still be with Cam cause she makes me really happy.

(II, NAS, 19)

Tim: I'm certainly not staying in the relationship for sex.

(II, NAS, 19)

Louisa: Like if there was no sex in the relationship you'd still be in the relationship?
Neil: Yeah, yeah, yeah. It's not a, not a big thing.

(II, NAS, 17)

The context within which this decentring of the importance of sexual intercourse took place was (as with the young women) partly attributable to the production of a 'safe' research environment. Most of these comments were offered in individual interviews where the absence of others meant their detrimental impact on young men's masculine identity was reduced. It is also likely that young men who volunteer to participate in research on sexuality possess a more flexible sense of sexual self. This is because talking seriously about sex and sexuality is not deemed an 'appropriately masculine' practice, and to do so may require a less rigid adherence to traditional perceptions of this identity.

In addition, *all* of the young men quoted here described themselves as currently in, or having previous experience of relationships. It is possible that the experience of being in a relationship influenced their perception of the importance of sexual intercourse in comparison with other aspects (for example, security, commitment, love, increased self-esteem and companionship[8]). Chris explained he stayed with Cam not because of sex, but because she 'makes me happy'. He later went on to describe, 'I was sitting in the park the other day with Cam and we were just messing around wasting hours and I just couldn't believe how good I was feeling, just being there with her'. Similarly, Peter who participated in the couple activity and focus group spoke during this session about what was so fulfilling about his relationship with Amy.

> Just being with somebody and knowing somebody just so well that, you know you can guess what they are thinking and what they are thinking all the time, it's just, yeah it's like, I feel like when we are together we are a whole person, when I am apart I am half a person.
>
> (FG, AS, 18)

The way in which experience of relationships may facilitate young men's access to less hegemonic constructions of male sexuality has been observed by other researchers. Holland et al. (1998) note that 'the development of an intimate relationship with a woman is a means by which many of them [young men] are able to distance themselves from the values of their male peers' (p. 90) thus enabling an experience of sexuality that is less rigidly constrained by the surveillance of the male peer group. Walker and Kushner (1997) also venture that having a girlfriend may have allowed young men to 'escape the ritualised banter of his mates and indulge in extended conversation' (p. 60). These findings indicate that for some young men (hetero)relationships offer a context in which less traditional expressions of male sexuality can be played out.

Young men's responses to an open-ended survey question which asked them to complete the sentence, 'what I want in a heterosexual relationship is …' also indicated many desired more than just sex from relationships. Responses could be coded into themes, with young men mentioning 'love' as the thing they most wanted in a heterosexual relationship. The next greatest number of mentions was 'trust, honesty, respect, followed by commitment'.[9] The flavour of some of these replies are captured in the following quotations where young men draw on what I refer to as a 'happily ever after discourse', where romantic notions of 'love', 'commitment', 'honesty' and 'caring' prevailed. Only those who reported engaging in sexual activity, (but not necessarily intercourse) and who described themselves as having had experience of relationships (although these may not have been enduring) were asked to complete this question. Their responses imply that experience of sexual relationships may affect young men's priorities about a relationship's benefits.

> Long term love and friendship. Someone to settle down with.
>
> (Q, AS, 19)

Having someone to love all the time and someone who loves you back. (Q, NAS, 19)

A caring, understanding, honest and loving relationship. Stuff that will make both people feel good while respecting their wants and needs. (Q, AS, 18)

Understanding, patience, being open and truthful about feelings. (NAS, 19)

Young men also referred to the importance of 'friendship', 'communication' and 'equality' within a relationship.

Intimacy, love, friends, partner. (AS, 18)

Communication. Sharing. (NAS, 19)

To be able to talk to each other and be truthful. (NAS, 17)

To be equal, romantic, fun loving. (AS, 17)

To have fun, be wanted, to enjoy myself, and be myself and be loved for that. (AS, 18)

While traditionally these sorts of relationship qualities have been documented as important to young women (McRobbie, 1991; Lees, 1993; Jackson, 2001) there is increasing evidence that this discourse is being taken up by young men. Interviewing young men aged 16–18 years about their experiences of heterosexual relationships Redman (2001) argues that romance was a 'common currency' through which subjects constituted their heterosexual masculinities. The importance men attribute to commitment in relationships is also evidenced in a British study of younger males (O'Donnell and Sharpe, 2000, p. 11). A majority of boys in this research agreed with the statement 'A wife and family of his own is the most satisfying thing that a man can have'. An increasing prevalence of these sorts of discourses associated with the concept of the 'new man' have been linked to changes in other social and economic practices. Edley and Wetherell (1997) characterise these as the 'feminisation' of the workforce and the shift from manufacturing to services industries organised around computer based technologies.

To suggest that this constitution of male (hetero)sexuality was the most constant in the research would however ignore the precarious nature of such conceptualisations and the ways in which masculinities are struggled over. In the following section I look at the complex and fluid ways in which young men (and women) are positioned within dominant discourses of male (hetero)sexuality.

Discursive manoeuvres; performing sexuality in the research context

As a 'modern technology' social research provides a performative space for the creation of versions of the self (Kehily, 2002). In the current study focus groups provided insight not only into what young people thought about sexuality, but their talk facilitated a 'performance' of their own sexual subjectivities. This environment was constitutive of young people's sexual identities by providing a public forum for the presentation of self. Through their talk about sexuality young people engaged in the management of their own sexual identities fashioning these through what they revealed and concealed about their sexual selves. This process can be seen as 'identity work' and encapsulates the idea that young people's sexual subjectivities are 'made' in the course of focus group discussion. This is a consequence of the collective interaction which characterises this method where participants discuss a chosen topic collaboratively rather than engaging in a two way conversation with the researcher (Barbour and Kitzinger, 1999). The public nature of focus groups (that is, the way in which talk is directed at an audience) and the fact that in this study participants were friends,[10] means they can function like a peer group. The role peers play in the enactment and regulation of particular cultures of masculinity and femininity has been well documented (Hey, 1997; Lees, 1993; Kehily and Nayak, 1997; Holland et al., 1993). Similar dynamics worked within the current focus groups engendering particular kinds of identity management amongst participants.

There is considerable complexity inherent in the concept of 'identity work' as it is not simply an accomplishment of will. The idea of young people's talk facilitating a 'performance' of their own sexual subjectivities means this is not an 'act' which they choose to put on. Rather it draws on Butler's (1993) proposal that gender is produced through the performance of repetitive acts which give the impression it is something constant which individuals possess. Identity-work is constrained by young people's access to culturally available resources which promote particular gendered versions of sexuality (Edley and Wetherell,

1997). In this way young people are both the subjects and objects of language in that their utilisation of it produces them as types of sexual subjects which are not of their own making (i.e. which they did not author). However the concept of identity-work also embraces the idea of what Willig (1999) calls 'the active and purposive uptake of subject positions by speakers' in discourse (p. 144). Willig explains this as the way 'individuals employ discursive constructions which afford positionings that help them meet objectives within particular social contexts' (p. 144). For instance, within the focus group setting young people can deploy discursive constructions with the object of preserving masculinity or projecting an appropriate femininity. This element of the conscious utilisation of language to manage the presentation of sexual self implies a level of self awareness not always present in general talk. It is this which distinguishes identity-work from the usual emergence of the subject through language.

Young men's identity-work in focus groups was intimately tied to managing vulnerability. If as Butler contends masculinity is a performance that can only be sustained through repetition it always 'resonates the echo of uncertainty' (Nayak and Kehily 1996, p. 227). This instability is because as a practice or set of practices it must constantly be achieved (Wetherell and Edley, 1998) leaving young men perpetually open to risk of failing to acquire a masculine status. The necessity to appear appropriately masculine and police transgressive male behaviour seemed to be the driving force behind the identity-work undertaken during discussion. What made this management of masculine identities particularly challenging was the way in which their purpose was a 'serious' discussion of sexuality, a task at odds with dominant expressions of masculinity. Haywood (1996) and Wood (1984) note young men's talk about sex is typically characterised by vociferous and public displays of sexism and bravado. In this way focus groups served precariously as a means of signifying masculinity and creating an environment in which masculinity's vulnerability was heightened. Some of this complexity is revealed in the following conversation about experiencing sexual pressure from girlfriends, a topic raised spontaneously[11] by several of the focus groups. In this extract a group of trainee farmers explain why some young men may not want to have sexual intercourse.

Louisa: What reasons would guys have for perhaps not wanting to?

Richard: The same reasons some girls don't want to may be they don't want to get AIDS or something like that. If they don't want to get them pregnant or......

Harry: Scared that it's not going to be good enough. *(others laugh here)*

Richard: Or if they don't do it right probably *(laugh)*.

Louisa: Is there like a pressure then for guys to feel like they have to uhm ......

Richard: Satisfy the girl or.....

Louisa: Yeah that they have to what's that term...perform? Is that a thing that guys feel?

Gail: Yeah.

Harry: Not me.

Richard: Oh you're just a studly *(laugh)*.

(FG, NAS, mixed)

This extract typifies many others in which young men projected a multiplicity of sexual selves, sliding from apprehensive and inexperienced sexual partner, to confident sexual predator in any one moment or conversation. This fluidity in the projection of sexual self occurs when Harry swiftly changes from declaring that young men may feel anxiety about not living up to their sexual partner's expectations to denial that this is a personal concern for him. While Harry makes a generalised statement about possible feelings of 'inadequacy' it is seen by other participants to reflect his own thoughts about himself. Their laughter signals that he has risked masculine identity and subsequently he hastily attempts to redeem a 'studly' persona by insisting he is not afflicted by such concerns. This identity-work occurs in direct response to the mocking reaction of his peers and might be understood as an example of the purposive deployment of discursive constructions to meet a specific objective (Willig, 1999). In this instance the discourses deployed are those which constitute a traditional masculine subjectivity in order to counteract the damage incurred by a less hegemonic constitution of self. This extract highlights the tensions young men may have felt between adhering to the demands of the focus group and answering questions about sexuality 'seriously', and the need to preserve an appropriately masculine identity. It also indicates the mutability of young men's sexual identities and ways in which in the course of identity-work they were constantly under modification.

The fluidity with which various discourses of masculinity are deployed and discarded in the presentation of the sexual self is also apparent in the following conversation between a group of young men at school.

Louisa: What does it mean to be committed for a guy in a girlfriend/boyfriend relationship?

Tawa: Ball and chain [Vaughn and Tem laugh]
Vaughn: Not that far but getting close to it.
Tawa: [Voice changes to serious] Oh being committed that's just
            uhm dedicating time aye to your girlfriend.
Vaughn: Yeah, yeah.

(FG, AS, mixed).

Tawa initially draws on a traditional discourse of male sexuality by implying his disinterest in commitment and lack of emotional investment in
relationships through the ball and chain metaphor. He engages in a denial
of emotional investment in relationships, a display which forms part of
a repetition necessary for the achievement of masculinity through its
concealment of (emotional) vulnerability. When Vaughn checks Tawa's
comment with the remark 'not that far', Tawa's tone of voice and attention change to a more serious vein, and he offers that commitment means
'making time' for a girlfriend. A softer version of masculinity is evoked
here, with Tawa giving recognition to the fact that relationships require
some input from him. It seems that Vaughn's statement invokes a less
hegemonic expression of male sexuality and is pivotal to Tawa's change in
tone and response to the question. Vaughn's reference to a 'softer' version
of masculinity appears to open the way for Tawa to undertake a discursive
manoeuvre in which he sides with his mate. Presenting a united front on
this issue and sustaining alliances within the peer group, structures the
identity-work Tawa engages in here. Such interactions shaped the versions
of sexual self which young men presented during group discussions and
indicate the complexities and fluidity in such linguistic posturing.

Young women's management of their sexual identities in focus
groups manifested itself in quite different ways and implies the gendered nature of identity-work. Like their male counterparts young
women's narratives served as a vehicle for the presentation of sexual
self during discussion and were constantly modified in response to the
comments and reactions of others. Hey (1997) and Skeggs (1997) have
described the way in which female friendship groups provide a context
for the enactment of particular cultures of femininity by demarcating
desirable and undesirable forms. In the process of agreeing with or
challenging statements made by others, young women collectively
defined feminine sexual identities and carved out their own. This
undertaking was structured by dominant meanings about being female
and sexual and what is acceptable for young women to do and feel.
Below, an all female group at school demonstrate how female sexual
identities are managed in this setting and some of the tensions young
women experience in this context.

Louisa: How do boyfriend/girlfriend relationships start?

Julie: Attraction.

Aroha: Yeah.

Eva: You go oh he's nice and then a friend will go 'I'll go and ask him out for you aye'? [laugh]. And you go 'no, no, no' and then they do it anyway [all laugh].

Becky: Well the first thing is definitely physical attraction although when we talk about it [pause] you know you always think you have to start off saying 'oh no I want my guy to be funny and kind and ....'

Julie: A nice person.

Becky: a nice person but that's never what we originally go for.....

Julie: Yeah.

Aroha: You go for their look, their appeal to you.

Julie: And if they've got brains it's an added bonus [all laugh].

Louisa: If they haven't got brains what happens then would you still carry on with the relationship?

Becky: You'd probably get quite bored, like if you were intelligent and the partner wasn't, the conversation would be lacking quite a bit because you wouldn't share sort of common interests or...

Julie: Once your relationship gets to a certain point, I mean the physical stuff is still there but you have got to have something more than just the physical attraction.

This kind of conversation was typical amongst all female groups in the research where participants portrayed sexual selves that offered a female equivalent of the male subject in Hollway's sexual drive discourse (see Chapter 6). The way in which young women prioritise male sexual attractiveness over 'being funny', 'kind' or having 'brains' mirrors traditional masculinity's preoccupation with physicality over personality. When Julie says: 'if they've got brains it's an added bonus' a conventionally masculine sexual subjectivity is constituted through the dismissal of the male subject as sex object. Given the potential risk to reputation which such assertions of sexual self carry (see earlier reference to Lees, 1993) it is interesting that they emerge in this context. The identity-work participants are engaging in here can be seen as a consequence of the presence of other young women and the atmosphere generated by this. In the absence of young men to claim these traditionally male sexual identities, young women are able to enjoy the sense of power they bestow and feed off the energy this invokes. In declaring that it is attraction that sparks heterosexual relationships a discursive space is opened in which both Aroha and Becky can challenge traditional discourses of heterosexuality and reveal themselves as desiring sexual subjects. Rich laughter evoked in this exchange is evidence of the delight young women procure from collectively expressing their experience of physical attraction.

Because a discourse of active female desire does not bear the same power as an equivalent positioning within the male drive discourse, it may not be sustaining for these young women. The threat it carries to their reputation may be too great, because unlike the male drive discourse it does not position them in a way that affirms their feminine status. After portraying a sexual self that deems physical attraction critically important in a relationship Julie, reverts to a conventionally feminine concern with the need for 'something more' for an enduring connection. Within the male drive discourse however, the ultimate relationship requires no other characteristics (like the need for a meeting of minds or interesting conversation). This shift in portrayal of sexual self occurs after her friend Becky also reverts to a more traditional constitution of female sexuality by commenting that she would 'get bored' if a relationship was based purely on physical attraction. In what might be conceived as conforming to the continual collective negotiation of feminine sexual identities, Julie responds to Becky's comment by affirming it and once again modifying her own presentation of sexual self. Like young men's identity-work in focus groups, young women's presentation of self is shaped by the reactions and responses of peers and thus perpetually under modification.

Concluding comments

Discourses which constitute youth as sexually innocent and any expression of their sexuality as 'deviant' or 'promiscuous' deny them a positive discursive space from which to be legitimately sexual. As sexuality is integral to humanity whether we are sexually active or not, denying this aspect of young people's self is potentially 'dangerous'. In the absence of more positive ways of understanding young people as sexual they are constituted as 'sexually innocent' or 'uncontrollable' both of which have negative implications for their sexual health. Offering young people spaces in which to perceive themselves as legitimately and positively seeking/experiencing sexual pleasure is likely to transcend some of the dilemmas posed by the 'gap' phenomenon. From such positionings empowered, responsible and satisfied sexual subjects can emerge.

Starting from the premise that young people are legitimate sexual subjects this chapter has sought to comprehend how they understand themselves as such. When young people talked about their sexual selves during
the individual and couple interviews this articulation was inextricably
tied to being 'sexually active'. This association was evident in the way
articulations of sexual subjectivity emerged most often in discussions
about engaging in sexual activity with a partner. I have suggested that
defining oneself as sexual only in relation to sexual activity with others
may limit how young women and men experience their sexuality. Power
relations which structure heterosexual relationships can reproduce inequitable gendered encounters and only encourage traditional expressions
of female and male sexuality.

In endeavouring to understand the social meaning young people
ascribe to their sexuality the chapter has analysed the hegemonic pressures from which sexual subjectivities are squeezed (Olesen, 2000). The
findings suggest that while young people draw on dominant discourses
of heterosexuality in the constitution of sexual subjectivity, their talk
involves a complex accommodation and rejection of these subject
positions. While this resistance is made possible through the constitutive force of discourse, such discourses may also be deployed as a
result of a subject's social location. Particular social locations may have
facilitated young people's access to, or opened space for, other ways of
constituting themselves as sexual. For example, in Shona and Anna's
case being in a supportive environment for teenage mothers, or for
some of the young men, having relationship experience.

While discourses of heterosexuality positioned young people as
particular sexual subjects there was also evidence of their 'active' and
'purposive' deployment of discourse to meet certain objectives within
the research context (Willig, 1999, p. 114). One objective within focus
groups was the management of their own sexual identities through
their talk about young people and sexuality. Through what it revealed
and concealed about their sexual selves this talk served as a vehicle for
the presentation of sexual self. For young men, such identity-work was
intimately tied to managing vulnerability and preserving an appropriately masculine status. While for young women discussion between
all-female participants provided a space in which to position themselves as active and desiring sexual subjects with a reduced fear of risk
to reputation. This identity-work was contingent upon collective affirmation of expressions of female sexuality as not 'sluttish' or morally
wrong. The fluid nature of this presentation of sexual self was evident
in the constant slippage of young people's talk as they purposively de-

ployed various discourses of heterosexuality in response to the reactions and comments of others.

These findings offer our exploration of the 'gap' equation and by implication sexuality education, an understanding of how some young people talk about their sexual selves. They suggest we need to acknowledge that young people's sexual subjectivities are nuanced and do not always neatly conform to traditional notions of passive female and active male (hetero)sexuality. Recognising this is important for the ways educational messages and strategies might tap into young people's sense of themselves as sexual subjects. For example, strategies which presuppose young women's lack of sexual desire or which ignore young men's aspirations for love may miss the mark, as they do not encapsulate some young people's conceptualisations of their sexuality.

The research also suggests that young people's constitution of sexual subjectivity is context bound. In the case of young men, their constitution of sexual subjectivity within a public environment and intimate relationship might vary appreciably. The fact that young men may not always take up the subject positions sustained by the operation of hegemonic masculinity could be a strategically useful tool for sexuality education. This ploy would only be effective however, if the pressure to appear appropriately masculine is also taken into account in educational messages and strategies. Shaping educational messages in ways that recognise the diversity and complexity of young people's conceptualisations of sexual self may in some cases, offer a more sophisticated approach to health promotion.

5
'Like I'm floating somewhere ten feet in the air': Experiencing the Sexual Body

Given that sexual practice typically involves the intimate engagement of bodies, it is interesting that traditional analyses of the 'gap' phenomenon have ignored corporeality. This omission has meant a lack of examination of how embodied aspects of young people's desire and pleasure may be implicated in this phenomenon. As a consequence the body has been an 'absent presence' in this work (Shilling, 1993), recognised implicitly as feelings and desires which propel young people into 'dangerous' sexual encounters. These disembodied analyses are reminiscent of the constitution of knowledge in sexuality education which attends to the body in an exclusively mechanistic and reproductive manner (see Chapter 3). One of the aims of this chapter then is to 'enflesh' thinking around the knowledge/practice 'gap' and consider how gendered bodies might be implicated in this equation.

This chapter is concerned with the way young people experience sexual embodiment or, how they describe embodied sexual experience. It engages with the criticism that studies of the body 'tend to suffer from theoreticism' confining their exploration to theory which is not grounded within empirical study (Nettleton and Watson, 1998). Taking account of this criticism the following exploration is based upon young people's talk about personal embodied sexual experiences, in particular, whether sexual embodiment is gendered, the forms this might take and the possible consequences for young women and men's sexual practice (i.e. whether this leads to pleasurable and positive experiences for them or not). Pleasure is highlighted here because it provides a measure of access to power by indicating whose pleasure is prioritised in social relations. Contribution is made to a general understanding of embodiment by extending this theorisation to the realm of young people's experience of their bodies within *sexual* practice. I formulate a politicised concept of sexual embodiment which provides a way of understanding how gendered bodies have access to particular configurations of power and experience positive/negative effects. By replacing the 'no body' in analyses of the knowledge/practice 'gap' with sexually embodied subjects, this chapter explores the implications of gendered sexual embodiment for comprehending this phenomenon.

Bodies of theory

Before embarking on an examination of young women and men's talk about (hetero)sexual embodiment it is necessary to understand how this discussion is located within current literature and debates about the body. Witz (2000) describes a 'corporeal turn' in sociology and feminist philosophy where there has been a 'recuperation of the body in/for social theory' (p. 2). In the last decade this has been evidenced by a proliferation of writing about corporeality (Holliday and Hassard, 2001; Gatens, 1996; Grosz, 1994; Weitz, 1998, Turner, 1996; Shilling, 1993; Scott and Morgan, 1993) and attributed to a culmination of factors such as; the rise of consumer culture, the advent of new technologies, the greying of populations and the politicisation of the body by feminists and those writing about disability (Turner, 1996; Shilling, 1993; Featherstone, 1991). The focus on sexual embodiment in this chapter forms part of this tide of interest in the body amongst those seeking a more complex picture of social life.

The concept of embodiment has emerged from the field of literature above to indicate 'the ways in which people live and experience their bodies' (Lupton, 1998, p. 82). Some writers differentiate between the notion of embodiment and 'the body' by arguing that the former entails more than simply possessing a bodily form. For Csordas (1990), embodiment implies something more than a material entity; 'it is rather a methodological field defined by perceptual experience and mode of presence and engagement in the world' (Csordas, 1994, p. 10). From this perspective the body is 'lived' and not just a fleshy shell which subjects inhabit. What is fascinating about the lived body is that in the course of everyday life its experience is typically taken for granted. A simple example is that as I think about this sentence I am largely oblivious to the work of my fingers on the keyboard and my eyes in reviewing these words. As Leder notes this state of 'hibernating' embodiment is characteristic of ordinary circumstances and is broken only when our bodies 'act up' as in the case of illness (Leder, 1990). While studies of embodiment have occurred around pregnancy (Earle,

2003; Longhurst, 2001), testicular cancer (Gurevich, Bishop, Bower, Malka, and Nyhof-Young, 2004), body-building (Bloor, Monaghan, Dobash, and Dobash, 1998), and teaching (Middleton, 1998) *sexual* embodiment has received minimal attention. Exploring how young people embody desire and pleasure within sexual relationships is a means of extending empirical and theoretical understandings around the body.

The concept of embodiment contains a central tension concerning the ontological status of the body and how it is experienced. This tension is explained by the ambiguous nature of embodiment which encompasses 'objective and subjective realities; nature and society; personal and interpersonal experience; interior and exterior states' (Watson, 2000, p. 110). In this way the body is perceived as existing as a 'body-for-itself' and as a 'body-in-itself' the former encompassing the 'natural', 'personal' and 'interior' aspects of embodiment while the latter refers to a social, interpersonal and exterior body. This conceptualisation of embodiment as encapsulating a 'living experiential body' and an 'exterior, and institutionalised body' (Turner, 1992, p. 41) mirrors the nurture/nature debate in which identity is understood as socially/discursively constituted or alternatively as a universal entity, existing independently of social context. To study sexual embodiment necessitates an explicitness about how embodiment and by implication the ontology of the body is understood and located within these debates.

In the current investigation the concept of embodiment draws variously upon the work of Grosz (1994), Foucault (1977), and Merleau-Ponty (1962). Unlike Cartesian approaches, in this study embodiment is seen as experienced by subjects who are ruled by a mind which exists in separate entity to a compliant body. Taking Merleau-Ponty's view, all human perception is embodied so that we cannot know anything independently of our bodies. This situation occurs because our bodies are the condition and context through which we are able to have a relation to objects. The world can only be known through the body's interaction with it, because it provides the situation or perspective through which information is received and meaning generated (Merleau-Ponty, 1962). As Grosz (1994) explains, it is 'sense-bestowing', and 'form giving', providing a 'structure, organisation, and ground within which objects are situated and against which the body subject is positioned' (p. 87). The mind and body are inextricably connected in a process of *being knowing* (as opposed to knowing and being) in a way that denies the Cartesian split. The importance of this conceptualisation of corporeality for understanding young people's experiences of sexuality is that it enables materiality to be taken into account and recognises the agency of the body as desiring, pleasure-feeling flesh.

Within this understanding embodied experiences cannot be extrapolated from the discursive world. This issue brings us back to the question of whether our experience of embodiment is biologically or socially determined. Some would argue that these experiences are the product of the way in which our flesh is inscribed by discourse (Malson, 1998) a perspective which maintains the nature/culture binary. This research takes the view that nature and culture are indeterminate as encapsulated in Grosz's (1994) description of the mobius strip.

> The Mobius strip has the advantage of showing the inflection of mind into body and body into mind, the ways in which, through a kind of twisting or inversion, one side becomes another. This model also provides a way of problematizing and rethinking the relations between the inside and the outside of the subject....the passage, vector, or uncontrollable drift of the inside into the outside and the outside into the inside. (p. xii)

Applying this conceptualisation to the experience of embodiment we might say that this experience is 'naturally social' (Grosz, 1987, p. 7). Discourses do not 'sit on the surface of the flesh or float about in the formless ether of the mind' but are instead 'enfolded into the very structures of our [bodily] desire in as much as desire itself is formed by the anonymous historical rules of discourse' (Stenberg, 2002, p. 61 my parentheses). As such our flesh *is* meaning, not simply for what it signifies (for example a gendered body) but because it is out of the indistinct entanglement of flesh and discourse that our subjectivity or as Merleau-Ponty would have it *knowledge* is made. For Foucault this is an effect of power about which he writes, 'nothing is more material, physical, corporeal than the exercise of power' (Foucault, 1980a, pp. 57–58). This theorisation renders our understandings of young people's sexual embodiment a product of circumstances that are 'naturally social'.

In this chapter I take a more politicised approach to the concept of embodiment than simply the 'ways in which people live and experience their bodies' (Lupton, 1998, p. 82). My concern is not just with how bodies are lived but the types of embodiment which offer positive (and negative) outcomes for young people. This interest rests on a premise generated by feminist thinkers that bodies have a gendered politics which produces qualitatively different bodily experiences for females and males (Weitz, 1998). Feminist thinkers have drawn on elements of embodiment theory employing concepts akin to Turner's 'body-for-itself' and 'body-in-itself' for the advancement of their own political projects. For example, Holland, Ramazanoglu, Sharpe, and Thomson (1994a) talk about how important it is for young women to recognise the difference between what Rich (cited in Fuss, 1989) has described as 'the body' and 'my body'. Empowering forms of embodiment are seen to eventuate for young women when they can distinguish between socially constructed expectations of how their bodies should be (i.e. 'the body') and the actual bodies they possess ('my body'). Inherent within this idea is a concept of the natural female body untouched by social context which young women can recognise/retrieve. Not only is this idea saturated with the well documented problems of essentialism (Brook, 1999) but it underestimates the indeterminacy of the effect of 'nature' and 'culture' on young women's embodied experience. If positive sexual embodiment were simply a case of cognitively recognising the social regulation of female bodies then body dissatisfaction would be passé. To further feminist analyses/politics what is needed is a theory that not only politicises sexual embodiment as gendered and unequal (in terms of positive experiences) but which encompasses an understanding of the complexity of experiences for a naturally social body. The following sections begin to develop this theory with reference to particular states of embodiment evident within young people's talk about their sexual bodies. The rest of the chapter's structure reflects the differential experience of sexual embodiment for each gender by dealing separately with young women's and men's references to these states.

Young women's narratives of sexual embodiment

Much of the writing around young women's sexual relationships has suggested that their experiences of sexual activity are mostly not satisfying and instead disappointing, especially in the case of inaugural sexual encounters (Hillier, Harrison, and Bowditch, 1999; Thompson, 1990). Feminists have explained this situation with reference to the operation of male power and the way this works to repress possibilities for young women to experience sexual pleasure outside of men's definition of it (Dworkin, 1987; Mackinnon, 1996). More popularised explanations have relied upon biological determinism suggesting that women's bodies are 'naturally' less adept at experiencing pleasure, an attitude epitomised by contemporary sex advice which suggests women need more stimulation than men to enjoy sex (Gray, 1995). These explanations invoke a subject who within feminist explanations is a product of her false consciousness (or mind) or a subject within popular sexology, whose biology is destiny. Whatever the reason, there is an evident disparity in young women and men's experiences of sexual pleasure with almost no literature suggesting young men's routine disappointment or lack of enjoyment with (hetero)-sexual encounters. Any theory of sexual embodiment which might offer young women (and young men) agency needs to acknowledge this discrepancy in gendered sexual experience. The development of a spectrum of embodiment is necessary to document a diversity of bodily experiences ranging from those that are positive and offer some form of agency to those that have negative effects. In this section we start at the positive end of the spectrum with what I have conceptualised as *embodiment*, a state which is largely positive and conveyed in young women's talk by an acknowledgement of their body's sensuality. 'Sensuality' pertains to the corporeal experience of pleasurable sensation induced by sexual activity.

Despite the prevalence of 'cultural stories' that constitute 'good girls' as not sexual (Tolman and Higgins, 1996) and which render talk about bodies and pleasures 'dangerous', many young women in this research provided examples of talk about embodied sexual pleasure. This 'embodiment' was witnessed in their positive acknowledgement of their bodies and the pleasures it offered them. In these narratives, young women made reference to *feeling* sexual pleasure and in some cases they isolated the bodily sensation which produced this experience. Below are two examples of talk revealing embodied female sexual pleasure during individual interviews. In the first, Emma not only articulates a sense of corporeal sexual pleasure but locates the region of pleasurable sensation in her body.

> Louisa: So how do you know what you find pleasurable sexually?
> Emma: It's like it feels really good in there [points to her lower abdomen] like I'm not here, like I'm floating somewhere ten feet in the air, that's when I sort of know that it feels nice.
>
> (II, AS, 17)

The next example reveals Ngaire's delight in the sensation of being physically close to her boyfriend describing this in flesh-filled imagery.

> Ngaire: ....it's just a feeling of oh wow this guy is lying next to me and I've got no clothes on and....it's just bodies twisted together.. uhm I feel close to him, I love that yeah. I think

that it is really nice to have someone you care about inside you, it's just really, it's amazing, it's beautiful *(laugh)*. (II, NAS, 18)

Embodied narratives were also produced in answers to a survey question that required young women to complete the sentence 'what I find pleasurable about sexual activity is...'

Ruth: The closeness involved, the feeling of naked skin together.
(Q, NAS, 19)

Helen: The feeling of touching someone and being touched.
(Q, AS, 19)

Lita: Foreplay, holding nakedness...
(Q, NAS, 19)

April: The feeling in my stomach and everything else disappears.
(Q, AS, 18)

Sandra: The feeling of his penis going in and out of me and him touching me all over and feeling me.
(Q, NAS, 19)

What makes these extracts appear embodied is the way young women describe pleasurable sexual sensations as lived that is, occurring in and through their bodies. The sense of the corporeal nature of this pleasure is evident in their depictions of the *feeling* of naked 'bodies twisted together', 'someone you care about inside of you' and a lower abdomen sensation that lifts you to the ceiling. These descriptions sit in stark contrast to notions of women's bodies as frigid or less likely to achieve pleasurable sexual response. Instead they describe bodily sensations that young women experience as both pleasurable and positive.

It is important to note that the narratives offered by young women in this research may or may not have any direct relationship with bodily feelings during sexual practice. Some young women may have experienced positive sexual embodiment but did not have the language, (or chose not), to articulate this. Similarly, other young women may have uttered the language of positive sexual embodiment, without having any lived experience of it. This possibility became apparent when one young woman explained that orgasms felt like 'tingling all over your body', yet later expressed confusion over whether she had experienced an orgasm.

Descriptions of female embodied sexual pleasure are available to young women through mediums of popular culture like women's magazines. Young women's reproduction of these in absence of a clear sense of their own corporeal experience suggests an interesting phenomenon which warrants further investigation. However, trying to 'verify' these experiences in young women's actual sexual practice is superfluous to my task here. It is in the fact that language constitutes them as a possibility/ reality for young women's corporeality that their potentially empowering and positive effects lie.

Young men's narratives of sexual embodiment

A conceptualisation of positive sexual embodiment for young men requires something more than their right to corporeal sexual pleasure as this is constituted as a condition of 'normal' male embodiment. Unlike young women, young men's bodies are seen as easily aroused and gratified an idea Tim conveys in this statement about his own experience of pleasure '....I mean if I wanted to ejaculate I could probably just do so in less than a minute' (II, NAS, 19). Despite purporting to be a description of sexual pleasure this comment is devoid of fleshy sensuality. Instead, it draws on dominant discourses of masculinity in which male bodies are constituted as machines that simply 'do sex' in order to achieve orgasm. Contained within this notion is the idea of a body that is detached from the sensuality of this experience. The key to young men's embodiment lies in transcending these dominant discourses about male bodies which may act as a constraint to a range of unconventional sensual potential. Embodiment as it is conceptualised for young men in this research, involves a recognition of the *sensuality* of the body in ways which diverge from traditional perceptions of male (hetero)sexuality.

Young men seemed to access this recognition of the sensuality of the body through emotional feelings. As a consequence of this process, sexually embodied narratives were more likely to be produced when talking about sexual experiences with partners young men were emotionally attached to. It was as if these emotional feelings elicited sensuality in their flesh, with several young men explaining that pleasurable corporeal sensation was intensified when they felt an emotional bond with their partners. In the following extracts young men illustrate this connection between emotion and increased bodily satisfaction through their comparisons of the experience of sex within casual and more enduring relationships. Chris talks here about how his first sexual experience with his girlfriend Cam was better than other casual sexual experiences he'd had due to the emotion he felt for her.

Louisa: So what was sex like the first time with Cam?

Chris: ...it went well, it was like astounding, wow it was good you know uhm *(laugh)*.

Louisa: Better than previous times?

Chris: Yes.

Louisa: So better than one-night stands?

Chris: It was better, it was just like it made sex worthwhile, like sex for me it wasn't worthwhile before, it didn't make you feel particularly good or anything and like sex with Cam is really good.

Louisa: So what's different?

Chris: I'm sure there must have been much more emotion in there, like there was real caring...like there wasn't....like this isn't because of sex, this was because, just because I feel good with her *(stabs finger into the table for definition)*.....Yeah sex for sex sake isn't worthwhile.

(II, NAS, 19)

Speaking about why sex with his girlfriend, Ngaire, was 'fun' and 'better' than a one-night stand, George explained:

> You know that you really know them, like before [with one-night stands] it was like turned on but you never get to know them or who they were or how they felt about such and such a thing...it's best if you can get to know them, that's the fun...the sex is better yeah.
>
> (II, NAS, 19).

Ashby described how sexual intercourse with his long-term girlfriend Becky, was more 'intimate' and 'comfortable' because of his feelings for her, whereas in a one-night stand sex for him was simply about lust:

> Uhm it's more intimate generally [in a long term relationship] uhm...yeah...yeah, yeah cause like I care about her and stuff like that...I'm just more comfortable with Becky *(small laugh)*. It's like uhm with the others it's just more lust yeah.
>
> (II, NAS, 19)

The link between emotions and a more pleasurable corporeal experience is apparent in each of these interviews. For Chris, the presence of 'much more emotion' in his relationship with his girlfriend made sex as he described it 'astounding'. Similarly, for Ashby, caring about Becky made his sexual experiences more 'intimate' and 'comfortable' than during one-night stands. Sex was 'better' for George because 'getting to know someone' is what produces 'the fun'. These narratives are embodied in the way that they diverge from traditional discourses of masculinity which construct male bodies as emotionally bereft and machine-like in their procurement of gratification. Instead, embodiment for these young men was revealed in the way that emotions offered a window to increased physical pleasure.

Embodied narratives also appeared in answers to a survey question where participants were asked to complete the sentence, 'what I find pleasurable about sexual activity is...'. Young men's embodiment was revealed in their reports of pleasurable corporeal feelings that were inextricably bound with emotional 'intimacy', 'closeness' and 'the thought of being with someone you love'. Below are some descriptions of what these young men found pleasurable about sexual activity:

The intimacy and the stimulation. (Q, NAS, 19)

Talking and being touched by my partner. (Q, AS, 17)

Close feeling, orgasm. (Q, AS, 17)

The presence (physical and mental) of a female. Being physically stimulated. (Q, AS, 18)

The intimacy and the pleasure. (Q, AS, 17)

Making love to the person that you love. Not just getting satisfied physically but the whole thing. (Q, NAS, 18)

Intercourse and oral sex. The thought of being with someone you love. (Q, AS, 18)

Here links between emotion ('the intimacy', 'love') and physical pleasure ('orgasm', 'getting satisfied physically') are made. As survey participants were anonymous the exact nature of this relationship could not be explored further. However, like the narratives offered above these comments suggest a state of embodiment in which the body is connected with emotions with the effect of intensifying pleasurable corporeal sensation. Lupton's work on the way emotions are seen as integral to notions of embodiment helps to clarify this phenomenon. She describes how emotions can be seen as 'embodied thoughts'; 'thoughts somehow "*felt*" in flushes, pulses, "movements" of our lives, minds, hearts, stomach, skin' (Rosaldo cited in Lupton, 1998, p. 82 my emphasis). This conceptualisation of a seamless intermingling of emotionality and corporeality makes it possible to understand how feelings of emotional attachment for a partner may manifest in more intensely pleasurable corporeal sexual sensations. While it is likely this experience is not confined to young men, the development of this notion of sexual embodiment requires a recognition of male sexuality as involving emotionality in a way that has typically been reserved for women.

Sexual disembodiment and dysembodiment

Two other corporeal states could be identified in the narratives of young people in this research, that of disembodiment and what I have conceptualised as dysembodiment. The idea of sexual disembodiment has been utilised by researchers from the Women Risk and AIDs Project (WRAP) in their examination of the sexual attitudes and behaviours of 150 young women in London and Manchester (Thomson and Scott, 1991). These researchers described many young women in their study as being disembodied, evident from the absence of talk about bodies and pleasurable corporeal sensations in interviews. Young women were 'disembodied in the sense of [experiencing a] detachment from their sensuality and alienation from their material bodies' (Holland et al., 1994a, p. 4). In the current research only young women's narratives appeared to display this state of disembodiment. This involved a 'corporeal sensual detachment' that was typically expressed by talk that lacked an acknowledgement of their bodies. Returning to my earlier comment about the need to recognise bodies as naturally social, this definition of disembodiment does not involve a splitting of 'I' from physiology. Disembodiment as it is used in this research is not the separation of the subject from their body, but in young women's case the body's lack of recognition of its sensuality, the body being inextricably cultural/biological.

While examples of embodiment emerged within the narratives of a few of the young women in this research, a sense of their disembodiment was more pervasive. Disembodiment was typically characterised by an absence of any reference to the body and/or pleasurable corporeal sensation in young women's talk about their sexual experiences. This kind of disembodiment has been well documented in other studies where young women speak (or rather do not speak) about sexuality as lived through the body (Tolman, 1994; Lees, 1993; Hillier, et al., 1999; Holland et al., 1994a). Such disembodiment was most noticeable within focus groups where a resounding silence about the body and female corporeal pleasure permeated interaction between many participants. This silence makes it difficult to provide illustration of young women's disembodiment as it was characterised by an absence rather than presence of talk. However, an example from one young woman during an individual interview will be discussed below. In terms of a notion of young women's sexual empowerment the state of disembodiment offered little in the way of pleasurable corporeal feeling.

Dysembodiment

To date theoretical explorations of the body have tended to polarise corporeal experiences into states of either embodiment or disembodiment. Within the course of this research it became obvious that these concepts did not capture the nuances encapsulated within young people's talk about their sexual bodies. For example, while bodies were 'missing' in some women's talk, others referred to a sense of lived-materiality but in a way which did not indicate a pleasurable corporeal experience. Rather than a lack of acknowledgement of the sexual body as in the case of disembodiment, these narratives demonstrated an acute awareness of the body produced through a sense of its perceived inadequacies. This bodily state was particularly evident in young women's descriptions during individual interviews when reporting what they were thinking and feeling about their bodies during sexual activity. These feelings were often negative, derived from anxiety that their bodies failed to emulate dominant ideals of feminine 'bodily beauty' and would subsequently turn their partner's off. As revealed below, such feelings of inadequacy had the capacity to impinge upon young women's experience of pleasure during sexual encounters. What sometimes crossed Becky's mind during sexual activity was that her boyfriend Ashby would think she was too fat and this thought made her feel 'ugly'.

Becky: ....it does affect me feeling fat and stuff cause sometimes when I'm on top and I can see my stomach you know and I hate that and my thighs you know, it's not sort of my upper body or uhm....

Louisa: And so what are you thinking about...do you feel about that at that particular time?

Becky: That he must just think that I look so grotts *(laugh)* and gross.

(II, 17, NAS)

Nina explained how she thought her boyfriend would be put off by seeing her naked, and that this made her feel self conscious during sexual intercourse.

> Nina: .....I used to be so self conscious...with like him seeing me with no clothes on and stuff...I just used to think I was really fat and stuff and that like if he saw me he would just be put off totally kind of thing.
>
> (II, NAS, 17).

Another young woman admitted she experienced 'fat days' when she felt dissatisfied with the size of isolated body parts. Although she remarked that such feelings did not influence the sexual activity she engaged in with her boyfriend, she conceded they had the potential to.

> Cam: Uhm I have a few hang ups about my body and stuff....
> Louisa: What kinds of things do you worry about?
> Cam: I have fat days. Just like the bloaty feelings and uhm my stomach and my buttocks and thighs and small breasts and stuff like that.
> Louisa: Do you think how you feel about your body affects sexual activity?
> Cam: It could if I let it...like wanting to keep like sheets on and stuff like that, to like cover things up like having the lights off all the time or something like that.
>
> (II, NAS, 19)

Amy disclosed that negative feelings about her body did affect sexual activity with her boyfriend Peter, whom she consequently tried to prevent from seeing her body.

> Amy: ....sometimes it does affect the sexual activity because I'll sort of be embarrassed and shameful....I'll sort of be a bit sort of like this [uses her hands to shield her body]. And he'll be it's okay get the arms away and I'm like I don't really want him to look at me.
>
> (II, NAS, 18)

In thinking about how these narratives might be conceptualised theoretically I turn to the work of Williams (1996) who utilises the term *dys*-embodiment to describe 'embodiment in a *dys*functional state'

(p. 23). He takes 'dys' from the Greek prefix signifying 'bad', 'hard' or 'ill' as it is present in words such as 'dysfunctional' and applies it to the bodily state chronically ill patients often encounter where '....the painful body emerges as "thing like", it betrays us and we may feel alienated and estranged from it as a consequence' (p. 27). Dysembodiment is an appropriate term to describe the way in which the young women above are aware of their bodies but in a *negative* mode. Such embodiment is *dys*-functional in that young women experience their bodies as 'bad' as a consequence of the discursive production of femininity where their failure to imitate cultural standards of bodily perfection invokes feelings of self-loathing and shame. Dysembodiment does not involve a distortion of a pre-existing form of female embodiment that is in some way more natural and positive. I would suggest that pleasurable corporeal feelings are no more 'natural' than dysembodiment, although they may be more useful in terms of developing a politically empowering conceptualisation of young women's bodily experience. In recognising the pleasurable corporeal outcomes of sexual activity, discursive space is made for revised meanings of traditional (hetero)sexual practices, that offer young women more positive outcomes.

The categorisation of sexual embodiment into three states implies that young women experience them statically. However, these states can be seen to lie on a continuum along which young women vacillate. This continuum ranges from narratives of disembodiment where the body and its pleasurable corporeal sensations are missing and moves to talk about dysembodiment where the body is acknowledged in a negative way. At the other end of this spectrum lies embodiment where young women's narratives reveal a positive awareness of corporeal pleasure. To elucidate how this continuum might operate, it is useful to concentrate on one of the most articulate participants Becky, whose narratives about her body reflect those of other young women in the study.

Becky had been going out with her boyfriend Ashby for just over three years and commented that sexual activity with him had not really been that pleasurable until recently. During the individual interview I asked her to talk about her bodily experience of this sexual pleasure. In response Becky declared she experienced orgasms (her interpretation of sexual pleasure) about sixty percent of the time during sexual activity. When pressed further about how this pleasure felt she divulged 'It's hard to say. Sometimes I don't even really know what's happening (laugh), you know it's like I'm not sure what's happening'. Her answer made no direct mention of her body or the feeling of orgasm and instead offered a clinical appraisal of pleasure in terms of percentages of orgasm. Becky's disembodiment is signified by the fact that although she is referring to a sexual sensation experienced corporeally, the sensual body is missing from her talk. This absence is something Becky may have partially recognised as indicated by her confusion about not being able to determine what she felt during sexual activity. While this moment of talk appeared disembodied, at other times Becky's narratives suggested dysembodiment as when she explained that '....sometimes like I look at myself while we are having sex and feel like absolutely huge and disgusting' (II, AS, 17). Becky's sense of how her body feels is clearly evident in her description of this sensation as 'huge' and 'disgusting'. The fact that these feelings are negative and not a source of pleasure for her render this talk about dysembodiment. What Becky's and other young women's narratives about their bodies suggested was that they oscillated across embodied states. Recognising this fluidity accounts for Becky's claim that around the time she was interviewed sexual activity was pleasurable and subsequently more embodied, yet at other points in the research her descriptions revealed dysembodiment or disembodiment. This vacillation across sexually embodied states expressed within young women's narratives enables us to acknowledge the diversity of bodily experience and potential agency contained within this conceptualisation of embodiment. The key is in understanding how and why movement across these sexually embodied states occurs.

Young men's narratives of dysembodiment

Lack of evidence of corporeal pleasure is a main condition of the way I have conceptualised young women's disembodiment. Within this research young men's talk did not suggest that bodily experiences of sexual pleasure were difficult to achieve or that they felt confused about what their bodies were experiencing. Typically, young men spoke of sexual pleasure as an experience that was easy and 'guaranteed' as indicated in Chris's comment about sexual intercourse '....a guy is sort of almost guaranteed to feel good you know, feel the same in the end anyway so' (II, NAS, 19). Like Tim's earlier cited statement about being able to 'ejaculate' if he wanted to 'in less than a minute', pleasure has a mechanical rather than sensual quality here. These descriptions are constituted through discourses which equate masculinity with 'sexual virility, potency and undeterred/unencumbered performance' so that such depictions of pleasure become a signifier of masculine identity (Gurevich et al., 2004). These narratives might be described as dysembodied in the way that possibilities for corporeal sexual pleasure are limited by dominant discourses of masculinity. In line with traditional notions of hard masculinity, in such talk pleasure is conceptualised as a mechanical and inevitable outcome of sexual activity rather than a sensual experience. Corporeal pleasure is therefore confined by what is perceived as appropriately masculine narrowing possibilities for young men of knowing other ways of experiencing pleasure.

The gendered nature of sexual dysembodiment is apparent in that for young women it is about a negative sense of the body produced through cultural ideals of female bodily beauty. For young men, dysembodiment suggests the confinement of their experience of corporeal pleasure to that which is deemed appropriate within traditional discourses of masculinity. Both forms of dysembodiment are intimately connected with the discursive production of femininities and masculinities and have the potential to restrict the experience of sexual pleasure. What differs is that this dysembodied state offers young women minimal agency while providing young men with some positive effects. While young women's negative sense of their bodies may lead to self loathing and the inhibition of pleasurable corporeal sensation, there is some status for young men in a machine-like body. This body not only delivers sexual pleasure regardless of its dysembodied state, but within the gender order male dysembodiment and authority share a close association. As Stenberg (2002) explains 'to admit to inhabiting bodies is to admit a weakness' (p. 54) so that young men's lack/denial of sensuality in their descriptions of sexual pleasure is an expression of power tied to the constitution of traditional masculinity.

The power associated with the state of male sexual dysembodiment is apparent in the following young men's narratives. A common occurrence when asked about how their feelings about their bodies might influence sexual activity, was for young men to deny being consciously aware of their bodies. Typical responses to this question were, 'Uhm, I don't know. As far as sex goes, I don't really think about my body', (II, AS, 17) suggesting their own body was not something they acknowledged or thought about in relation to sexual activity. Other young men accompanied their statements about their lack of bodily awareness with comments which suggested their disinterest and dismissal of it as important or relevant in sexual situations. This sentiment is captured in Neil and Chris's use of the phrase 'I just don't care' below.

> Neil: No, no I don't feel, I just don't feel anything about it, sort of
>   over cocky you know 'Oh I've got a good body' so I just

>   don't care. My body's just there and it doesn't worry me you
>   know.
>
>   (II, NAS, 17)

>   Chris: I don't think that it matters. I mean I don't really care what
>   my body looks like. I do care what Cam [his girlfriend]
>   thinks my body looks like.
>
>   (II, NAS, 19)

This refusal to contemplate the body as worthy of consideration has been noted by Grogan and Richards (2002) in their work around young men and body-image. They report young men perceived body-image as 'trivial' and the devotion of time to thinking about or actually changing their bodies is conceptualised as 'female-appropriate behaviour'. These researchers explain this reaction as unsurprising 'since bodily concern is a stereotypically feminine concern and runs contrary to prevailing ideals of masculinity where body function (rather than aesthetics) is valued' (p. 228). Such bodily disassociation might have been conceptualised as evidence of disembodiment in young men's narratives, however careful examination of the interview transcripts uncovered other narratives which sat in tension with these comments.

Despite alleging when asked directly, that they didn't think and didn't care about their bodies all of the young men who participated in the interviews revealed that they did think about them and that these thoughts centred around bodily anxieties. In these instances young men spoke about their body's perceived 'inadequacies' referring to having 'small biceps', 'lack of muscle' or not being 'toned' or being 'short'. These statements indicated their comparison of their own bodies with a socially constituted ideal male body generally 'characterised by a well developed chest, rippling arm muscles and a "washboard stomach"'. The following extracts come from individual and couple interviews with men who were currently in a relationship. Chris, referring to his girlfriend seeing his body, described how his insecurities meant that 'some days I just don't, I just don't want her to see me' (II, NAS, 19). Further into his interview Chris explained that this was because:

>   Just uhm a feeling of inadequacy because, like you know like I feel
>   that, I feel that Cam [his girlfriend] is so amazing and that I'm not
>   really good enough, like that's just like a I dunno it's probably irratio-
>   nal but you know it's like you really, really, really want to make the

other person happy and like if you don't have say Carlos Spencer's[1] body *(laugh)*...Cam has a huge crush on Carlos Spencer.

(II, NAS, 19)

Indicating some feelings of insecurity about how attractive his girl-friend found his body, Neil described how he thought she wanted him to be more stereotypically masculine:

I mean sure there's different things that you want on everyone. I mean I am sure Nina wishes that I was a bit more muscular or what ever else...

(CA, NAS, 17)

After saying he didn't care about his body, Ashby revealed his concern about being 'too short', and the feeling that he had to compensate for this with extra muscle in order to 'look good' and be a good sportsman.

Louisa: So going to the gym and doing weights and stuff like that is to make you feel better or...why do you do it?

Ashby: Really it's for sports but uhm it does have, it does make you look better and stuff like that. But uhm my basic uhm motivation uhm is just for sport cos I'm a bit on the short side so I've got to, you know, I've got to try and make myself bigger.

(II, AS, 17)

Research by Magdala (2002) suggests that although men are generally more satisfied with their bodies than women, body dissatisfaction and related problems such as eating difficulties and anabolic steroid use are on the increase. The above narratives attest to body image anxiety experienced by some young men and their sense of 'failure' to emulate a muscular 'ideal'. Magdala notes the media plays a central role in the dissemination of this standard of male attractiveness which has recently become even more muscular. Evidence of this is seen in the increasing muscularity of boys' action figures, the contemporary popularity of weight training machines, gym memberships and performance-enhancing nutritional supplements as well as the visibility of muscle on programmes such as the World Wrestling Federation matches which draw their largest TV audience from young men (Magdala, 2002).

The fact that some young men thought or worried about their bodies, even though when questioned directly they maintained that they did not, might be explained by the operation of hegemonic masculinity. Hegemonic masculinity involves young men denying or protecting their own 'vulnerability', where vulnerability entails their failure to achieve dominant perceptions of masculinity. For young men in this study, admitting that their bodies were 'inadequate' meant exposing their vulnerability. Within the operation of hegemonic masculinity successful men are constituted as 'invincible', and have strong imposing bodies that reflect this image. As Connell (1987) writes, 'what it means to be masculine is quite literally to embody force, to embody competence' (p. 27) something communicated through the display of muscle. Referring to their bodies (except in relation to sport) or drawing attention to their worries about them, may have threatened to sabotage these young men's achievement of an appropriately masculine identity.

Thinking about how this phenomenon might be placed within a theory of sexual embodiment, it is apparent that this is not the same kind of disembodiment depicted in young women's narratives. In young women's talk about sexual experiences bodies are *missing*. As young men did make eventual reference to corporeality, it was rather that during direct questioning about their bodies these were *denied*. When these bodies are finally invoked young men's talk about them in negative terms suggests dysembodiment. Like young women's narratives around dysembodiment there is a suggestion that this state is at least partially the consequence of dominant discourses of bodily perfection. While in young women's talk this dysembodied state appeared to have the potential to effect their experiences of pleasure, it did not seem to extend such power over male pleasure.

Implications of a continuum of embodiment for the 'gap' equation

Having established the gendered nature of young people's sexual embodiment I now want to consider the implications of this for thinking through the 'gap' phenomenon. Hillier et al. (1999) argue that when young women lack a sense of embodied sexuality or available language to speak about the body, this can have important ramifications for their sexual health. 'In order to practice autonomy and agency in sexual encounters women need to feel connected to their own bodies and have access to language to express their needs' (p. 83). In the current research young women's disembodiment characterised by a silence around the body and confusion over what if anything their bodies were feeling, is suggestive of an absence of autonomy. The way in which young women's bodies are missing from their narratives implies a lack of awareness about what they are feeling and doing. This state may see them engaging in behaviours they have not positively assented to or not had an active role in creating. Examples here, might be engaging in sexual behaviour they do not find pleasurable or not having thought/or been able to suggest the use of a condom. Although possessing a sense of agency may not ensure safer sex practices occur, at the very least this is the starting point from which they might. Thinking about how the continuum of sexual embodiment set out here might indicate whether knowledge gained in sexuality education will be applied in practice, it seems that the state of disembodiment is unlikely to engender this.

Young women's narratives of dysembodiment which demonstrate a negative awareness of their bodies as failing to live up to cultural standards of bodily beauty may also mitigate against the translation of sexuality education's messages into practice. Positive body image is related to self esteem, something which young people need if they are to recognise they are worth protecting from sexually transmissible diseases, abusive sexual relationships and in order to enjoy sexual experiences. As several of the young people in this sample acknowledged, feeling bad about their bodies could potentially prevent this sexual activity from being a positive and pleasurable experience. In addition, because dysembodiment is intimately tied to dominant discourses of femininity that foster young women's passivity, this may work against young women as active agents in sexual encounters.

For young men, dysembodiment suggests a different set of gendered impediments to practicing knowledge absorbed from sexuality education. If young men's dysembodiment is informed by traditional discourses of hard masculinity which constitute a disassociation from the body, then this may also have negative consequences for safer sex practice. A body that is machine-like in sexual encounters and which is disconnected from feelings of sensuality and sensitivity is not likely to be concerned with the potentially negative consequences of sexual intercourse. The self-gratifying pursuit of pleasure and disregard for other needs and emotions (its own and possibly those of a partner) would seem to deter such concerns. This state may also cut young men off from a sexual experience that is more sensually pleasurable than the traditionally mechanical male body is 'allowed' to feel.

As an embodied state that may be most conducive for the application of knowledge into practice sexual embodiment appears to hold the most promise. Its potential lies in the way it involves a positive and pleasurable embodied experience during sexual activity. As pleasure is not simply a natural consequence of corporeality but indeterminately linked with gender politics, achieving pleasure within sexual relations suggests young women exercise agency within these encounters. If we return to Hillier et al.'s (1999) point above about the importance of agency in securing young women's sexual health, it is the narratives of young women's embodiment which offer the greatest possibility. Similarly, young men's narratives of embodiment hold the most hope of the implementation of safer sexual practices. These practices are more likely when embodiment is accessed through emotional attachment to a partner where caring about their sexual health may come into play. As will be discussed in Chapter 6, Moore and Rosenthal (1998) and Waldby, Kippax, and Crawford's (1993) work suggests that emotional attachment to a partner with whom some sort of enduring relationship has been forged may impede safer sex practice. However, I would suggest that this finding has untapped potential for sexual health promoters wanting to capitalise on the issue of love in relationships in a positive way with campaigns that assert 'If you love me, protect us and wear a condom'. One of the issues with considering the implications of these gendered states of embodiment for the 'gap' phenomenon is that they were fluid. Young women and men's talk appeared to slide across these embodied states revealing for instance, sexual embodiment at one moment and dysembodiment at another. This vacillation suggests that states of embodiment are perpetually in flux and their implications for putting knowledge into practice constantly shifting. Being able to elucidate the contexts and contingencies of these states would be a valuable undertaking.

Conclusion

This chapter has attempted to literally add flesh to the 'gap' equation by considering how bodies are implicated within this phenomenon. As part of this task it has sought to contribute to embodiment theory by drawing on young people's narratives of their sexual experiences to understand how the body is lived in this context.

The empirical data offered here suggests that the way in which young women and men embody sexuality is gendered. This embodiment is mediated by the discursive production of femininity and masculinity producing differential access to power and pleasure for each. For example, talk revealing young men's dysembodiment appears to consolidate male power and offer them guaranteed sexual satisfaction.

In contrast, young women's narratives of dysembodiment indicate a self-loathing of their bodies which acts as a potential impediment to their sexual pleasure.

To avoid the pitfalls of biological essentialism it has been important to recognise this gendered sexual embodiment as the consequence of a naturally social body. This means recognising that sexually embodied states are not simply the product of natural biology or discursive constitution, but an indeterminate entanglement of flesh and discourse. What is gained from this conceptualisation is an acknowledgement of the complex way in which power and pleasure are embodied.

One of the main aims of this analysis has been to develop a politicised notion of sexual embodiment which not only reveals its gendered nature but which opens spaces for agency. This task has involved problematising the conceptualisation of these narratives as either embodied or disembodied and proposing the addition of another category named dysembodiment. It is in recognising the fluidity of these embodied states as they appear in young people's talk, that possibilities for gendered sexual empowerment occur. For young women, this might include a more positive sense of their bodies and pleasurable sexual experiences in which they exercise agency. Similarly, young men might transcend the confines of hard masculinity that narrow their options for sexual expression and experiencing a corporeal sensuality that is more satisfying.

A final element of this discussion has been to speculate on the implications of this spectrum of sexual embodiment for the application of knowledge acquired within sexuality education in practice. While the states of disembodiment and dysembodiment may mitigate against teachings such as safer sex, sexual embodiment provides a platform from which this practice may be more likely to occur. In the next chapter we examine the way in which sexual embodiment is lived in the context of relationships.

6

Desire, Pleasure, Power: Understanding Young People's Sexual Relationships

Chapter 4 revealed that relationships are the context in which young people's sense of themselves as a sexual person is realised and formed. As such relationships are an important site for contextualising our understandings of how young people live out (hetero)sexuality. Relationships are also the place in which sexual knowledge can find application and exploring their micro-politics may shed light on why knowledge retained from sexuality education is not always translated into practice. The aim of this chapter then, is to gain a sense of relationships as experienced by young people in this research. A number of key questions are addressed in order to better understand how relationships offer a site in which young people's knowledge and subjectivities are played out. The first question is descriptive and asks what do young people's relationships look like in terms of their length, frequency and way they are characterised for example, as fun or serious. How young people talk about the pleasure they derive from relationships is also considered. Endeavouring to understand how pleasure is conceptualised in this context is a political act in its acknowledgement that this is an important element of young people's lives and one often institutionally ignored. A final question concerns the micro practices of relationships and the gendered operation of power within them. It asks how do young people describe decision making and negotiation around sexual activity in their relationships?

Before embarking on an exploration of the these data, I want to make a general point about the nature of young people's relationships. It is commonly presumed that the relationships young people engage in are taken less seriously by them and involve less personal investment than those that may follow in later life. Relationships during youth are often considered 'immature' and lacking the commitment and depth of emotion which characterises 'adult' intimacies. Such perceptions are drawn from discourses which constitute adolescence as a time of fickle desires reflecting a need to 'play the field' before selecting and settling into the right relationship (Coleman, 1980). As a derivative of ideas which align young people's sexuality with 'deviance' and 'promiscuity' this construction implies an adult rather than youthful perspective. The fact that half of the couples who participated in the activity and individual interview had been going out for more than two years, suggests some young people have enduring relationships where a particular level of commitment and personal investment is implied. In fact, the transcripts are peppered with examples of the deep feelings and importance young people assigned these relationships in their lives.

As a means of continuing the exploration of young people's sexuality through their own articulations, this chapter examines young people's relationships as depicted by them. It is structured in four sections, the first documenting the sexual relationships young people described in terms of number, length and type. These demographics provide a backdrop to ensuing discussion about the micro-politics and practices of these relationships. Given that it is often the negative consequences of young people's relationships which are emphasised (such as unplanned pregnancy and STIs) the next section explores relationship pleasures and how these are conceptualised. To determine how young women and men experience their sexual subjectivities within a relationship context, the next section examines experiences of negotiating sexual activity. The way in which power is sexualised and the implications for how each gender might experience their sexual subjectivities is also analysed. The last part of the chapter holds particular significance for understanding the 'gap' phenomenon in the way it explores how young people perceive their knowledge as affecting their sexual practice. It builds on earlier findings in Chapter 3 by describing *knowledge in practice* from the viewpoint of young people themselves.

The couple context: young people's (hetero)sexual practices

While there is a substantial body of international literature on young people's (hetero)sexual relationships we know comparatively little about the intimate details of these in New Zealand.[1] National based studies which interrogate the intricacies of heterosexual relationships have tended to focus on older age groups[2] instead. This section aims to fill this hiatus by creating a picture of the samples' relationship history and providing some insight into their experiences in this realm. As this kind of information was most efficiently collected in the questionnaire, findings from this method are reported here. Such data provides important information for people who work with youth around sexuality and those involved in sexual health promotion, design and delivery. Knowing young people's relationship status, whether they are engaging in sexual activity or not, with whom and under what conditions may help to target sexual health messages more effectively. Gaining a sense of how young people conceptualise their relationships and what they report as transpiring within them, provides a clearer picture of areas health campaigns might target and possible strategies to employ.

Findings from the questionnaire revealed that the majority of participants (50%) perceived themselves as having had a relationship at the time of the questionnaire distribution, while 41 per cent were currently in one. These results indicated their interest in these partnerships and that most young people felt they had relationship experience of some description. To preserve young people's own conceptualisation of what a relationship was, no time frames were specified to determine relationship status (e.g. longer than a week). Subsequently, whatever young people constituted as a relationship counted, whether this involved a matter of hours or years. In their research with a younger population of rural women in the West of England, Morris and Fuller (1999) observed diversity in the conceptualisation of relationships. Characterised as a continuum, these ranged from '....the transitory, "getting off", to the more serious "seeing someone" to the very serious "going out with", a longer term relationship more closely resembling adult partnerships' (Morris and Fuller, 1999, p. 537). Findings from the qualitative data suggest that young people in the current study 'counted' similar types of relationships as part of their experience. Young people talked about relationships as 'one night stands' which involved physical interaction of some description ranging from kissing only, to engaging in sexual intercourse. These relationships mirror those of 'getting off' in Morris and Fuller's study in their often single event character. The possible inclusion of sexual intercourse (which was not mentioned as part of the repertoire of sexual activities Morris and Fuller's participants engaged in), may be a reflection of an older age groups' sexual behaviours. Participants in the current study conceptualised 'seeing someone' and 'going out' like those in Morris and Fuller's research, as indicated in the following focus group discussion between young women at school.

Louisa: So what kinds of relationships are there out there then?
Dina: Casual dating sort of thing?

Louisa: So is casual dating different from what you were talking about before, you know the one night stand thing?

Dina: Well one night stand, you sort of know that, that is a one night stand, and that's it and long term relationship you sort of know that, but when you are casually dating you don't really know where it is going, you don't...

Rowena: You are unsure where you stand with the person. If it is supposed to be a relationship or if it's just a friendship with a few perks to it. (*The others laugh at this.*)

Sally: There's people that you don't consider yourself to be going out with but (*pause*) maybe if you go to a party and you get with them sort of.

Dina: Then that can get really confusing cause one of them might think, one partner in the relationship might think that you are together and then the other one might think that you are not.

(FG, AS, mixed)

These young women make a distinction between what they conceptualise as 'casual dating' and a 'long term relationship'. Although this terminology reflects New Zealand youth culture, these relationships share similar features to 'seeing someone' and 'going out' in Morris and Fuller's study. While with casual dating there may not be a sense of oneself as a permanent 'couple' there is some regularity with which sexual activity phrased by Sally as 'getting with them', is engaged in. Talk within other focus groups implies that this sexual activity or the 'perks' to which Rowena refers, may comprise petting or intercourse. This discussion indicates that 'casual dating' involves considerable ambiguity in the way it suggests a more regular relationship than a one night stand, yet less time invested than in a long term relationship. As Dina points out, its wavering nature can sometimes lead to confusion between parties about how the relationship is understood. Such talk during the qualitative methods helped identify types of relationships young people may have been envisaging when answering the survey question about how many relationships they had experienced.

Of those who reported never having been in a relationship, young people who described themselves as religious were most likely to be in this category. This result may be accounted for by the fact that many religions prohibit relationships of a potentially sexual nature prior to marriage. Gender analysis of these data disclosed that more young men reported they had been in a relationship while more young women described themselves as currently in one.[3] Being in a relationship is not a necessary marker of masculinity if young men can demonstrate a string of past exploits which prove their 'pulling power'. Even in the absence of an audience like the male peer group (the questionnaire was an individual project and anonymous) the performative nature of gender requires young men enact a version of (hetero)sexuality that is associated with desirable masculinity. The repetitive power of these performances may be evidenced here in the way they are invoked without the physical presence of others.

The average age of first dating experience for both young women and men was reported as 13 years old. The range of first dating age was wide however, with subjects reporting this anywhere between 5 and 19 years old. These data were collected from a question that asked 'Approximately what age were you when you started going out with boyfriends/girl-friends'? As described above, young people used their own definition of what constituted a 'girlfriend/boyfriend', so that all varieties of a relation-ship, not just those that were stable, long term or committed were incorporated. The inclusion of a concept that is deliberately not defined by the researcher, and therefore might be variously understood by participants would be considered meaningless by traditional statisticians. However, this questionnaire sought to document young people's own sense of what relationships were regardless of their diversity, something this question achieved by allowing participants to decide what relationships to include.

When asked how many boyfriends or girlfriends they had experi-enced, both young women and men reported an average of 5.8.[4] A small minority, six young women and four young men, reported 20 or more partners. While these numbers may appear high the question asked about *partners generally*, and not just *sexual partners* as has been the tendency of other studies primarily focused on sexual behaviour (Dickson, Paul, and Herbison, 1993; Carroll, Volk, and Hyde, 1985; Holibar, 1992 p. 199). This concentration on *sexual* partners is often a consequence of the health promotion motives underpinning such research and a subsequent need to identify risk. Research on sexual relationships of this age group in New Zealand indicates for example, that young women's mean number of *sexual partners* is 1.5, while young men's is 1.7 (Dickson et al., 1993 p. 199). These numbers are much lower than those sourced in the current study and may indicate that of the 5.8 partners reported, most may not have been partners they had experienced sexual intercourse with.

A majority of young people who answered the questionnaire stated that they went out with partners who were a mixture of ages, or with those who were older than themselves. Young women were more likely to go out with older partners[5] while young men were more likely to date young women the same age or younger than themselves.[6] This trend is reflected in other international and New Zealand based survey research (Dickson, Paul, Herbison, and Silva, 1998) with this age group. For example, American research involving 122 women in their twenties revealed that their first sexual partners were significantly older than those of the young men in the sample (Guggino and Ponzetti, 1997). There is a common perception that many young women prefer to date men older than themselves because their faster physical and emotional maturation renders boys their own age less attractive. Morris and Fuller's (1999) study also suggests that there is some 'status' and material benefits for young women in going out with someone older. Older partners are more likely to have an income, access to a car and patronise bars and pubs that offer young women (especially rural young women) a level of 'freedom' that they might not otherwise experience.

The length of young people's relationships is a detail which is seldom systematically recorded, possibly because there is a sense that at this age the frequency and brevity of relationships would make this difficult. Findings from this study reveal a majority of young people reported their current relationship being between 6 and 12 months while 30 per cent indicated that their longest relationship had been between 1 and 5 years. These figures suggest some stability in young people's partnering, contesting the view this age group's relationships are generally brief. These data also demonstrate that young women were more likely to have been in their current relationship for 1–2 years while young men for only 1–3 months.[7] This finding resonates with discourses which Hollway (1984, 1989) identifies as influential in constructing accounts of heterosexual relations in the West. The first of these is known as the 'male sexual drive discourse' which describes male sexuality as motivated by a natural, biological sexual drive which men must satiate. Another is the 'have/hold discourse' which configures sexual relationships within a romantic ideal where concepts of emotion, connection and commitment preside. In relation to these discourses the questionnaire results reveal the way young women's reports of 'enduring relationships' position them within the have/hold discourse as concerned with long-term partnership. Read within the 'male drive discourse' young men's shorter relationship length can be seen as symptomatic of a need to pursue the male sexual drive. Analysing these results through such discourses also ties young men's reports of more previous relationships (rather than stability in a current partnership) to the importance of consolidating masculinity by appearing sexually experienced.

A majority of participants reported that their shortest relationship fell within the one week to one month time frame indicating diversity in the range of relationship length. These sorts of relationships may mirror the terms of 'getting off' and 'seeing someone' in Morris and Fuller's research where pairing off as a one time or more frequent event (but which did not involve spending time alone doing things as a couple) were observed. Interestingly, there were no significant gender differences between the number of young women and men who reported engaging in one night stands. The positioning of young men within dominant discourses of heterosexuality as emotionally remote and sexually voracious constitutes them as more likely to have one night stands. This is a pattern of sexual behaviour which previous research has documented.[8] More recently, and as the current research serves to highlight contemporary (hetero)sexualities appear to be more flexible in their transgressions of gendered sexual boundaries. Stewart (1999) for instance notes how in her research women aged 17–18 years revealed a clear rejection of traditional passive femininity in their delight of 'picking up' on Friday nights (Stewart, 1999, p. 279). Similarly, Harris, Aapola and Gonick (2000) draw attention to the way in which young women in Australia, Finland and Canada appeared to negotiate the sexual double standard. These findings and those from the current research, suggest that the social landscape of gendered sexual identities may be shifting so the thrill of 'getting off' or a one night stand is also openly enjoyed by young women.

In terms of the occurrence of sexual activity within young peoples' relationships, 88 per cent of the sample reported that they had engaged in sexual activity. Jackson (1984) has argued that a 'coital imperative' operates within heterosexual relations conflating sex with sexual intercourse and making sexual activity without vaginal/penile intercourse unthinkable. Given the benefits of non-penetrative sex in combating the transmission of HIV there may be considerable value in decentring the importance of 'sex' in young people's relationships. In an attempt to disrupt the 'coital imperative' and encourage a more wide ranging conceptualisation of sexual activity, the questionnaire inquired about sexual behaviour broadly. It asked participants if they had engaged in 'sexual activity' defining this as '...engaging in petting (hugging, kissing) or sexual intercourse with someone'. This meant all forms of sexual activity (not just sexual intercourse) were counted.

However, in other research with 18 year olds in New Zealand 58 per cent of young men and 68 per cent of young women reported sexual intercourse in the previous 12 months (Dickson, 1996, p. 227). Read in relation to the current research these findings suggest a large proportion of the sample may have experienced sexual intercourse.

A consequence of this conceptualisation of sexual activity was that it highlighted the power of the coital imperative and young people's investments in it. During feedback about the experience of completing the survey one young woman expressed her dislike of being seen as sexually active under the terms sexual activity had been defined. She commented that she did not feel that having engaged in 'just kissing and that' made her sexually active and she wanted to resist being seen as such. This young woman's comment might be understood as a desire to preserve her status as sexually innocent given its social currency in particular settings. Being known as a 'good girl' i.e. not sexually active, engenders particular benefits in environments such as the school and family where such conduct is rewarded in the form of being considered a bright student or cherished daughter. Concurrently, this young woman's feelings convey the power of discourses which conflate sexual activity with sex by indicating the investment subjects can have in them. When 'not being a slut' and 'being a good girl' are reputations that rest on virginal status the imperative to sustain the demarcation between foreplay and sexual intercourse becomes evident.

Pleasure and desire in relationships

Young people's relationships are often associated with negative outcomes such as sexually transmissible infections, unplanned pregnancy, sexual coercion and date rape. The need to avert these sorts of consequences underpins the philosophy of many sex education programmes which draw young people's attention to them as a means of prevention. Media also play an important role in constructing teenage relationships negatively in the sensation value they reap from publishing stories such as 'Eleven year old girl gives birth to 14 year old boyfriend's baby', 'Ten year old boy has sexual relationship with his teacher'. As a consequence, there is a paucity of talk about pleasure and desire as a positive experience within young people's relationships. In addition, a global climate in which cases of HIV/AIDS augment daily has the effect of framing research on young people's relationships in terms of discovering information which might signal strategies to minimise risk of this disease. While references to the pleasures young people gain from relationships surface in some of this work (Holland, Ramazanoglu, Sharpe, and Thomson, 1998; Tolman, 2002), in others they are left unacknowledged or downplayed to findings about unequal gendered power relations and their consequences for safer sex negotiation (Brander, 1991). This relegation is partly because the pursuit of pleasure and experience of desire are seen as arch enemies of sexual health promotion, when in the 'heat of the moment' the condom is forgotten. I want to argue however, that pleasure and desire might be seen as potential vehicles for understanding and combating the 'gap phenomenon'. This argument will be taken up in more detail in the next chapter while here some foundations are laid by acknowledging what young people find pleasurable about their relationships and how they experience desire within them.

As indicated earlier, documenting such practices is a political act which extends the argument made in Chapter 4 about recognising young people as sexual subjects. This stance acknowledges that seeking and expressing desire and pleasure are legitimate aspects of young people's sexuality. By discussing young people's sexual relationships in direct relation to pleasure, I attempt to reflect participants' interest in this area of sexual knowledge. As Chapter 3 indicated, participants felt that knowledge communicated through sexuality education was removed from their interests and as such, held lower status than knowledge obtained through practical experience. What young people were specifically interested in was a discourse of erotics in which information about desire and pleasure were integral and eagerly sought through discussion with friends and/or from sexual practice. By describing what young people portray as pleasurable about relationships and how they experience and express desire within them, what is offered is empirical evidence that might inform a discourse of erotics.

Pleasure and sexual practice

Data on the sexual activities young people engage in during relationships and what they found pleasurable about these could be found in the 'Pleasure Sheet' completed by six couples who participated in the couple interviews. While one partner was with me during an individual interview, the other filled out a sheet naming particular sexual activities they had engaged in during their present and previous relationships and whether or not they had found these pleasurable. The activities ranged from hugging/kissing, mutual masturbation to anal sex and sharing sexual fantasies. Space was also provided for subjects to indicate any activities not specified (however only one subject utilised this). From these sheets the most common sexual activities engaged in were hugging, kissing, sexual intercourse, oral sex, mutual masturbation, sexual touching and sharing sexual fantasies. Although not referred to by other subjects, one young woman who had been in her relationship for 9.5 months added that she also enjoyed 'talking dirty'. The least practiced activity was 'anal sex' with only one young woman having engaged in this during a previous relationship. This echoes findings from an Australian study of 18–19 year olds, where only five per cent of (hetero)-sexual couples reported this practice (Rodden et al., 1996). Most of the activities were found to be pleasurable by all subjects, although two young men indicated that they did not find giving a partner oral sex so.

In order to establish whether young people found sexual activity pleasurable or not, the questionnaire asked subjects to 'agree' or 'disagree' with the statement that 'Sexual activity is pleasurable'. This statement was one of 15 items aimed at determining young people's attitudes towards sexual activity. Of those who responded to this question 92 per cent of young women and 94 per cent of young men agreed that sexual activity was pleasurable indicating a resounding majority of participants had lived experience of pleasure, (or if they were not sexually active thought that this activity could be pleasurable).

Another open ended survey question asked subjects to complete the sentence, 'What I find pleasurable about sexual activity is...'. Only those who described themselves as having been sexually active[9] with a partner answered this question. Answers revealed an array of corporeal and emotional pleasures derived from (hetero)sexual activity which could be coded into a series of themes. The most consistently mentioned answers referred to a general notion of 'pleasure and enjoyment' and made reference to this activity being 'fun' and 'enjoyable' or as one young man described it 'exhilerating' (sic). The following is a sample of representative answers to this question with original grammar and spelling preserved.

Enjoying being with my partner. (AS, 17, Female)

I love the fun of it. Very exciting. (NAS, 17, Male)

That I can have fun and do what I want to do in a fun manor. (AS, 17, Female)

You get enjoyment from both sides. (AS, 17, Female)

The level of enjoyment I get from it. (AS, 18, Male)

As this question was posed within the context of an anonymous questionnaire there was not an opportunity to define exactly what subjects found 'fun' and 'exciting' about this activity. However, other answers to this question discussed below give some additional insights. Young women and men's answers to this question made equal reference to this theme suggesting that the fun and enjoyable aspect of this activity was an equally relevant pleasure for both.

The feeling of 'togetherness and closeness' sexual activity can generate was another popular specification for both genders. This 'togetherness' was expressed in terms of the way physical proximity incited a pleasurable mix of corporeal and emotional sensation. These statements were differentiated from others that could be categorised under the theme 'pleasurableness of emotional intimacy' in that they made specific reference to the way physical closeness increased emotional bonds. The statements below reveal how being physically close to a partner was seen to evoke pleasurable emotional feelings.

Feeling you get when you are close to them. (AS, 17, Male)

Being close to someone I love. (NAS, 17, Female)

The feeling of being so close to someone. (AS, 17, Male)

The feeling of closeness and affection between two people. (NAS, 18, Female)

The intimacy and the feeling and the bond afterwards. (NAS, 19, Female)

More young women than young men reported that physical closeness incited positive emotional feelings.[10] These responses generally drew upon dominant discourses of (hetero)sexuality where young women are constituted as having greater investment in the emotional merits of relationships, while young men are perceived to be more concerned with its corporeal benefits (Duncombe and Marsden, 1993). Holland, Ramazanoglu, and Sharpe's (1993) research has revealed how sexuality is a central site in men's struggles to become successfully masculine and this involves a disengagement with so called 'feminine concern with emotion'. On the other hand, the expression of young women's sexuality is regulated by their need to manage their sexual reputations (Lees, 1993; Holland, Ramazanoglu, Scott, Sharpe, and Thomson, 1991; Thomson and Scott, 1991). Displaying too much interest in the physical pleasures of relationships (without emotional investment) puts young women in danger of

gaining a negative sexual reputation. Given these disciplinary features regulating young women and men's (hetero)sexualities, it is unsurprising that
more young women than men wrote about 'togetherness and closeness'
being a pleasure gleaned from sexual activity.

Answers which indicated what young people rated as the third
most pleasurable thing about sexual activity were coded under
the heading 'positive feelings associated with the body'. These
expressions of pleasure were grounded in bodily sensations which
produced a positive and enjoyable corporeal experience. This
acknowledgement of pleasurable corporeal feeling was evident in the
following statements.

The feeling of having my partner inside my body. (AS, 17, Female)

The closeness involved and the feeling of naked skin together.
(NAS, 19, Female)

The feeling of the penis going in and out of me and him touching
me all over and feeling me. (NAS, 17, Female)

My partner's body against mine. (NAS, 18, Male)

Feeling, kissing, staring, hugging. (NAS, 17, Male)

Slow touching, and getting hot/sweaty. (AS, 18, Male)

There were no differences between the number of young women and
men who named positive feelings associated with the body as pleasurable. However, when these themes were ranked by gender, young men
named 'positive feelings associated with the body' as the second most
pleasurable aspect of sexual activity, while for young women this came
third after 'togetherness and closeness'. While positive emotional
rewards featured in young people's answers to this question (as seen in
the themes of 'increasing emotional intimacy' and 'pleasurableness of
emotional intimacy') on their own these did not rate as highly as pleasures of the 'flesh'. Some caution should be exercised in reading these
results as young people finding the physical benefits of sexual activity
more pleasurable than their emotional advantages. In responses to this
question, it was difficult to distinguish whether participants were referring to the emotional or physical pleasures of sexual activity, because
in many answers these were inextricably entwined. This interweaving

was clearly seen in the first two themes of 'pleasure and enjoyment' and 'togetherness/closeness', which may be indeterminately derived from physical/emotional sources. One particularly interesting result is that a gender difference was found in answers which described 'emotional intimacy' as being a pleasurable aspect of sexual activity. In a manner that appears to disrupt dominant (hetero)sexual subjectivities, more young men than young women indicated they found this pleasurable. This finding lends support to the fluid nature of masculinities as discussed in Chapter 4.

Experiencing desire

There are few studies which have documented young people's experience of sexual desire especially within the New Zealand context (Tolman, 1994).[11] This study sought to determine if young people perceived themselves as desiring subjects and to what extent. In something akin to a question from a magazine 'sex survey', subjects were asked, 'How would you describe your sexual desire'? The options provided were 'Very strong', 'Average', 'Weak' and 'None'. The majority of subjects (57%) described their sexual desire as average, with more subjects likely to say their sexual desire was 'strong' rather than 'weak'. While more young men said their sexual desire was 'very strong', young women were more likely to say their sexual desire was 'weak'. However, these differences were not significant. The fact that young people in this study rated their sexual desire as neither 'rampant' nor 'weak' but instead 'average' may imply (and other New Zealand research would suggest) that strong desire is not always the reason for their engagement in sexual activity (Dickson et al., 1998).

Exploring how young people expressed their desires was the focus of another survey question. The intention here was to disclose how sexual desires were given expression in a relationship with a well known partner. In response to, 'How would you express your sexual desires to a partner you knew well' subjects could tick one of the following options:

a) By telling them.
b) By showing them what I like and want.
c) By telling and showing them what I like.
d) I don't express my sexual desires to my partner.

The majority of subjects indicated they would both 'tell and show' their desires to a partner with the next preferred method being to simply 'tell them'. A significant gender difference was found here with more young women indicating they would do this. That young women would take an additional step and physically act on their desire, by showing a partner what they liked, is an interesting contravention of dominant notions of female sexual passivity. As Kenway and Willis (1997) reveal, young women's sexual subjectivities are the result of a mutable set of resistances and accommodations which depend upon contextual factors and access to power (p. 65). The fact that showing someone what they liked is less likely to be an option that young men adopt may be a result of increasing education around sexual harassment and prohibitions around 'touch', especially men touching women. The influence of these messages might also be seen in significantly more young men than women disclosing that they 'do not express their desires to their partners.'

Negotiating sexual activity in relationships

Having examined what young people's relationships look like and the pleasures they derive from them, this section is concerned with their micro-politics, specifically the way in which sexual activity is negotiated. The need for negotiation around sexual activity was apparent from young people's descriptions of points of tension and disagreement in their relationships. Issues of contention were most likely to be where and how often to have sex and the positions and types of sexual activity engaged in. When participants in the couple activity indicated how they negotiated these conflicts they explained that communication and being respectful of the other person's desires helped them resolve these situations. They also indicated that some decisions were made at the outset of sexual activity and then remained unspoken throughout the relationship's progression, tending to be automatically assumed rather than negotiated in each instance. These forms of negotiation have been explored in more depth elsewhere (Allen, 2003) and generally suggest a distribution of power between partners which draws on a discourse of 'equality'. Negotiation where communication is paramount and a mutual feeling of satisfaction over decisions is achieved represents an ideal of power relations. From their narratives, it appeared most of the young couples wished their relationships operated in this way and implied that at various moments they did. While all of the young couples expressed this sense of decisions being made equally, for brevity only one example is quoted here:

Cam and Chris described their sexual decision making as a mutual activity with no one person assuming control.

> Cam: I reckon it's....I reckon it's equal it's like we decide equally, like when and how often and how.
>
> Chris: Yeah see the decisions yeah it's not just decisions about sexual activity we make together it's like everything. Cause there is no one in charge here.
>
> (CA, NAS, 19)

Heterosexuality has often been theorised by feminists as a repressive institution in which men exercise power over women (Kitzinger and Wilkinson, 1994). While heterosexuality can operate as an oppressive institution, its power is not monolithic and neither are young women and men 'docile' subjects who act in complete absence of agency within its nexus. The above narratives indicate young people were able to perceive their sexual relationships as involving power sharing and not simply the designation of power to the male by virtue of his sex/gender. At least at a perceptual level, young couples in this study saw institutional forms of heterosexual power as contestable.

While couples constituted power relations as 'equal', it was apparent that sometimes young people's narratives of sexual negotiation involved a mediation of power. This was particularly the case in relation to how sexual pleasure was experienced in the relationship and by whom. Within prevailing discourses of heterosexuality it is men's needs and desires which are prioritised reflecting unequal gendered power relations (Hollway, 1984). While young women produced narratives which appeared to affirm this situation, they did so in a way that seemed to maintain their sense of agency. For example, Ngaire and Becky presented their prioritisation of their partner's sexual pleasure over their own gratification as an active choice.

> Ngaire: If I don't, if I don't give him one [an orgasm], I get upset and think, oh we've got to do it again until you do *(laugh)*...all the time I want to and uhm I don't mind if I'm not satisfied as long as he is.
>
> (II, NAS, 18)

> Becky: ...I wanted him to get what he wanted from it you know, and I didn't really care cause I was quite happy just for him to you know [orgasm] and sometimes he'd say 'oh do you

want me to stop' and I'd be like 'oh it's okay' even though I sort of wanted him to, I still wanted him to get what he wanted from it. Because I didn't want him to be dissatisfied.....

(FG, AS, 17)

On one level these narratives appear to comply with constructions of heterosexuality that give preference to the greater importance of male pleasure and the requisite of male sexual gratification. However, they also invoke a sense that men's pleasure and sexual needs take priority because these two women permit it. Agency is implied in the way that Ngaire says she is upset when her partner doesn't have an orgasm and *wants* to try again so he can, and when Becky explains that she wanted her partner 'to get what he wanted from it'. It seems that these young women have reconstituted their own pleasure so that it is indistinguishable from that of their partners.

For other young women who participated in the couple activity, privileging their partner's pleasure over their own was not stated as explicitly as Becky and Ngaire. Evidence that they did this was glimpsed when I asked Cam about satisfaction and pleasure in her relationship and she spoke about wanting to make sex more pleasurable for her partner Chris, 'I feel like that in some ways I could make it better for him and like I've got to try and find a way to make it better. I don't know why I feel that way it's just how it is' (II, NAS, 19). Later in the interview she expressed dissatisfaction with the way Chris sometimes rushed the part of sexual activity she described as 'foreplay', 'Chris's really good with stuff like that but other times he's not and...it doesn't mean that the sex is bad it's just not as fulfilling as it could be' (II, NAS, 19). Despite her dissatisfaction she had never mentioned this to Chris because she found '....it hard to say like...things like that to him because he's very sensitive about stuff like that anyway that he's not satisfying me enough and I feel like if I said it to him...he'd get very I don't know on guard' (II, NAS, 19). Instead of confronting Chris with her own lack of satisfaction, Cam pushed this aside to assert she needed to try harder at pleasing Chris sexually, clearly prioritising his sexual gratification over her own. Similarly, Nina described in her individual interview how she sometimes faked orgasm with her boyfriend Neil. She explained this in terms of wanting to give Neil the impression she had enjoyed sex and that he had pleased her – 'I have, I have faked it but that is only to make him feel better'. In faking orgasm, rather than demanding her own corporeal satisfaction from the relationship Nina symboli-cally concedes Neil's pleasure is more important than her own (Holland et al., 1998; Roberts, Kippax, Waldby, and Crawford, 1995). However, Nina describes the act of 'faking' as something she decides to do out of her own desire to 'make him feel better'.

In these instances, young women reported a kind of mediation of male power by actively 'choosing' to assent to it. They claimed they placed their own sexual pleasure in secondary importance, in order to enable their partner's needs, pleasures and sense of appropriate mas-culinity to take precedence. In a sense, they appeared to actively subject themselves to this power, seeing it as something they had control over through their active participation in sustaining it. They constituted themselves as exercising power because they had *made a choice* to continue with sexual activity they weren't really enjoying, or to give their partner's physical pleasure in the absence of their own. However, such exercises of power were limited by the fact that these contexts were largely defined by young men's pleasures and needs.

The operation of power in the extracts above might be theorised as an example of effective patriarchy whereby young women have been duped into servicing male requirements to their own disadvantage. It might also be argued that they are subject to a disciplinary power which produces them as 'docile bodies' complicit in the process of their own subordination (Foucault, 1977). Both these analyses would seem to negate the sense of agency with which young women describe actively making the decision to put their partner's pleasure first. To say that a mediation of power was occurring here is not to deny power's disciplinary effects, but to suggest that the subject produced is more than 'docile' or totally determined by this power.

In addition to young people's narrative portrayals of 'equal' and 'mediated' power, a final type – coercive power, was evident. This oper-ation of power involved young men endeavouring to compel partners to engage in sexual activity they did not want or were not sure about, through verbal or physical means. Coercive power was not demon-strated by couples during their interaction in the research context. As might be expected the 'rules' governing the research situation served as a restraint on abusive behaviour. Inconsistencies across narratives pro-duced in different methods indicated the equal distribution of power young couples recounted, was not always a reality (all of the time) in their relationships. This disaccord was revealed in discrepancies between their purported opinions about coercive sexual behaviour and other narratives which depicted actual behaviour in relationships. Without exception, during the couple activity all subjects emphasised, that if either of them did not want to have sex or did not want to perform a particular activity they would not be impelled to do so.

Ashby: ....if they really don't want to do it then you know there's nothing you can do about it. You can't force them or anything.

(CA, AS, 17)

Chris: Well uhm...if someone doesn't want to do something you can't force them to do anything I mean you can't I mean it would just ruin what we have between us if I was to force my opinion on her.

(CA, NAS, 19)

Neil: Cause if one of us doesn't want to have sex we won't we'll just say 'no'.
Nina: It's not like.
Neil: You don't feel obliged.
Nina: Yeah we don't have this thing where I will because you want to, but I don't really want to kind of thing.

(CA, NAS, 17)

Peter: Cause it has happened before one person wants it and the other person doesn't or can't or something like that.
Louisa: So what usually happens in that kind of situation?
Amy: Then the person who doesn't usually wins out, or, I mean if you were to go ahead it would be rape really and that's not you know that's not something that Peter and I want to go through.

(CA, AS, 18)

While resolute in voicing these opinions about their relationships, evidence from their partners or other narratives they offered indicated unequal power relations were sometimes or predominately at work. In Nina and Neil's case decisions about sexual activity appeared to not always be decided as equitably as portrayed. Nina described a period in their relationship when Neil wanted her to perform oral sex on him and she refused. At this time Neil used physical force in an attempt to pressure her into this activity. Nina explained, 'He'd always keep on pushing my head down there and I'd go "ow don't do that"' (II, NAS, 17).

Clearly, Neil used his physical might to exert power over Nina to engage in an activity she did not want, indicating that sexual decision making was not always conducted as democratically as depicted.

Similarly, although Ashby insisted on the importance of not 'forcing' someone to have sex, Becky his girlfriend spoke in a focus group session about continuing to have sex when she didn't really want to. Adding support to this, Ashby hinted that he may have sometimes coerced Becky into sexual activity:

> You know if like you can try and persuade them but you know *(Becky laughs)*. You can say 'Are you sure, are you sure' you know I mean, you know give them a taste...
>
> (CA, AS, 17)

Chris, (who also stated above that sexual activity should be an equal choice), admitted in the individual interview the first time he had sexual intercourse with his girlfriend Cam, he had pressured her into it, 'I didn't want to, I didn't want to put, I mean like I was putting a bit of pressure on her like.....'. That Cam was reluctant to have sex was communicated during the individual interview when talking about her emotional anxieties surrounding the decision to have sex (see Chapter 4) and the way she described her experience of this sex as 'indifferent'.

Amy also revealed that although Peter had never forced her to have sex when she had said no, he certainly exerted pressure on her. This was disclosed in two separate comments she made during the couple activity, as follows:

> Amy: Like uhm he might be in the mood but I might not be and it's sort of like you know 'oh you never want to give me any' *(putting on Peter's voice)*.
>
> (CA, AS, 17)

> Amy: I mean Peter will usually say 'How about it' or something like that, it sort of you know 'oh no, I don't really feel like it tonight' or you go *(to Peter)* 'Oh come on, please man'.
>
> (CA, AS, 17)

Clearly, Peter did not take Amy's 'no' as her final answer and in some instances appeared to have exerted additional emotional pressure 'you never want to give me any' to try and persuade her into sexual intercourse. While in Nina and Neil's relationship the expression of power was one of overt physical force, the last three examples demonstrate more subtle forms of coercion. These involved emotional pressure and persuasion through sexual enticements which Ashby described by revealing how he tried 'to turn Becky on' to sexual activity by giving her 'a taste' of how good it would be for her. These examples are differentiated from the operation of mediated power in the way that young women had expressly stated their desire not to engage in a particular sexual practice. In the examples of mediated power above, young women did not articulate their reluctance around particular activities instead making a private decision to participate in an activity that subordinated their own physical sexual pleasure.

What is apparent here is a fracturing between young people's construction of negotiation and its contingent power relations and their descriptions of their practice, a phenomenon they may not have been consciously aware of. Young people's depiction of a relationship in which power is always shared equally, may have been attributed to a kind of 'wish fulfilment' in absence of this situation in reality. The likelihood of this was referred to by one young woman during a focus group. In a general discussion about young people's relationships she commented that it was common to 'say oh yeah I've got an equal relationship but you don't, no one ever does I don't think' (FG, AS, 17). The fact that admitting imbalances of power may invoke unpleasant feelings could be a reason why couples did not dwell on this in their talk.

Other New Zealand survey research supports the finding that young people experience sexual coercion in relationships with 18 per cent of subjects aged 16 and over reporting at some time being forced to have sexual intercourse (Coggan, Disley, Patternson, and Norton, 1997). Clearly, sexual coercion and violence feature in young people's experience of sexual relationships. However, I would argue that dismissing the exercise of power that young women describe as experiencing around sexual negotiation in relationships, runs the risk of ignoring the complexity with which heterosexual power operates. This sense of power to participate in sexual negotiation apparent within the conceptualisation of 'equal' and 'mediated' power is at the very least experienced at a conceptual level by these young women (Holland et al., 1998). There was additional evidence in the young women's talk that in some instances this might have been achieved at an experiential level also. For instance, after breaking up with Ashby because he had slept with someone else, Becky accepted him back on the basis he abided by rules she had set concerning the use of contraception, having an HIV/AIDS test and telling her where he was going and what he was doing. He had managed to adhere to all such stipulations up until the time of their participation in the research.

Similarly, Neil and Nina's relationship had undergone substantial changes since Neil had nearly lost his life,[12] which had prompted him to stop drinking and reflect upon his aggression. In addition, he had lost interest in sex while Nina was now the one likely to initiate this activity. In Chris and Cam's case, Cam appeared to have considerable hold over Chris by virtue of his love for her. In addition, she was planning a working holiday alone abroad which suggested an assertion of her independence.

All of these circumstances reveal that these young women exercised some power in their relationships and that their experience of male power was not simply one of domination. They were however also subject to more oppressive forms of power by way of pressure and coercion exerted by partners. These incidences reveal the stable exercise of oppressive male power in their relationships. Young women's portrayal of decision making as equal and their sense of being actively involved in this, suggests that heterosexual power incorporates some agency for them. As young women in this study describe themselves as having experience of power and provide evidence of its effects, this cannot be simply dismissed as 'false power'.

This interplay of power and agency within heterosexual relations may be what sustains the pervasiveness of male power. As Foucault (1983) notes, 'Power is exercised only over free subjects, and only insofar as they are free' (p. 221). Such thinking does not infer that the pervasiveness of male power is immutable or that the potential spaces for women's agency are pre-determined or finite. This would dismiss the gains which some women have made within the social order and their personal relationships. Rather, I would suggest that within heterosexual relations male power by its nature operates so as to always offer spaces for female agency, the potential extent of which is constantly shifting. This potential is governed by multifarious factors such as a person's social location (including their access to particular discursive resources) and the material and historical conditions in which they live. What these findings show is how the power configured by these circumstances is played out in some young people's conceptualisations of sexual decision making.

Knowledge 'in' practice

In accordance with the research's concern with re-examining the knowledge/practice 'gap' this section concentrates on how young people conceptualise the effect sexual knowledge has on their sexual relationships. In Chapter 3 it was noted that a majority of the questionnaire sample (60%) reported that their sexual knowledge did not effect their sexual practice. This answer was in response to the question 'Do you think that your level of sexual knowledge has affected your relationships/or ability to have relationships'? In this section I analyse the thoughts of the other 40 per cent who perceived an association between their sexual knowledge and relationship experience. These findings suggest that young people view their sexual knowledge as having a range of positive and negative relationship effects. The fact that not all participants saw sexual knowledge as having a positive impact, challenges sexuality education's premise that 'knowledge' (of a particular type) is implicitly good and will produce positive effects. While the majority of subjects reported experiences which resonated with this idea, for others such knowledge was described as damaging relationships by instilling a sense of apprehension and fear.

Positive effects of sexual knowledge

One of the limitations of the open-ended question above was that it did not always invite responses which indicated what young people thought their level of sexual knowledge was. This constraint meant determining if participants perceived a 'low' or 'high' level of sexual knowledge to produce the named effect was not consistently possible. Generally, participants' answers implied that they had interpreted the question in terms of their possession (rather than lack) of sexual knowledge and its subsequent impact on their relationships. This reading was evident in the following responses where sexual knowledge served as a guide for determining whether sexual activity would occur. In these extracts participants describe how this knowledge enabled them to know what sexual activity they did (and did not) want to undertake in relationships.

Know how far you want to go with someone. (Male, AS, 18)

I know what I want and I feel that I can be sexually active with my partner because of it. (AS, 19, Female)

It has affected our relationship because we know how to wait, so we are not having sex. (AS, 17, Female)

I didn't have sex until I was ready. (AS, 17, Female)

Providing young people with knowledge and skills that will enable them to make responsible decisions about their sexual behaviour forms the essence of many sexuality education programmes in New Zealand (Ineson, 1996). Even within the most conservative programmes, decision making around sexual activity constitutes a focus with abstinence promulgated as the 'right' option. Given that much sexual knowledge from 'official sources' is packaged in this way, it was predictable that young people would feel it clarified their sexual decision making.

In addition, sexual knowledge was also seen to have a positive effect on how sexual subjectivities were expressed in relationships. For some, it provided a sense of autonomy in sexual situations allowing them to see themselves as sexual actors. Participants described how having knowledge about sexual matters imbued them with confidence producing a feeling of control in sexual situations. Sometimes this self-assurance had additional relationship benefits, facilitating communication between partners who felt more competent to express their desires and concerns.

> Made me more confident, able to relate to partner. (AS, 17, Male)

> I'm more confident of myself, and that makes me and my partner feel good. (AS, 18, Female)

> You feel more at ease knowing things. Know how to handle a situation. (NAS, 18, Female)

> I am able to feel confident enough to be honest, truthful and talk freely about anything at all. (NAS, 19, Male)

> You know more so you're more confident in what you are doing. (NAS, 17, Female)

More confident sexual subjectivities and their by-product of being at 'ease' in relationships also appeared to heighten the pleasures some young people gained from sexual activity. This consequence was described by young women, who saw their knowledge as increasing not only the pleasure they experienced from sexual activity, but their partners' also. Part of this effect could be attributed to the way confidence gained from possessing sexual knowledge facilitated young women's ability to voice their preferences and desires around sexual activity:

> I understand how to give pleasure and how to talk about what I enjoy. (As, 17, Female)

It has helped my partners and I enjoy our sexual activities. (AS, 18, Female)

Well knowing about sex has helped my sexual relationships know about positions and stuff has helped me and my partners enjoy sex. (AS, 18, Female)

At the moment my partner and I have a very intimate and fulfilling sexual relationship and he was a virgin (practically) so I like to think my knowledge did a lot to help us get where we are. I'm very open about sexuality issues. (NAS, 18, Female)

I am not so insecure with myself and I am a lot more confident about saying what I like and what I don't like instead of just going with the flow and thinking this is how it is always done. (AS, 18, Female)

It is difficult to know without further questioning, which conceptualisation of sexual knowledge (i.e. the distinction established in Chapter 3 between that gained from experience or information gleaned from secondary sources) that young people perceived to cause these effects. As 'how to give pleasure' and 'positions and stuff' do not comprise the traditional content of sex education programmes it is likely that such knowledge has been obtained from other secondary sources or personal experience. These findings emphasise the importance of providing young people (and perhaps particularly young women) with knowledge about 'how to talk about what I enjoy' and insight into the logistics of sexual activity so that they might experience themselves as empowered in sexual situations. Sexual encounters where young people feel able to ask for what they want because they are confident in their knowledge of sexual activity, are far more conducive to safer sexual practices than an environment of sexual ignorance.

In their references to giving and receiving pleasure from a partner, the young women above draw on what has been identified as 'a discourse of reciprocity' within heterosexual relationships (Braun, Gavey, and McPhillips, 2003). This discourse refers to an 'ideal' of heterosexual encounters in which both partner's desires and needs are met and culminate in an encounter of mutual sexual satisfaction. While this kind of exchange appears to challenge conventional discourses of (hetero)sexuality where men's pleasure is privileged over women's it may also be potentially problematic (Braun et al., 2003). Problems lie in the way such a discourse may function 'in conjunction with other discourses, to promote certain "norms" for heterosex' (Braun et al., 2003, p. 241). Norms such as a coital imperative in which sexual activity is conflated with vagina-penis intercourse and an obligation for women to reciprocate the pleasure they receive regardless of their wish to do so. The fact that knowledge provides some young women with the confidence to request activities which will give them pleasure, may offer a more empowering experience of (hetero)-sex. However, the presence of a discourse of reciprocity contains the possibility that attached to this request for pleasure is an expectation that the 'favour' be returned. This discourse poses a threat to young women's right to refuse to engage in any type of sexual activity.

Another positive consequence participants reported was that having sexual knowledge heightened their awareness of the possible dangers of sexual activity. This outcome was perceived as beneficial because knowing about sexually transmitted infections and other negative repercussions of sexual intercourse could assist in avoiding these. This benefit was demonstrated in the following responses where young people talk about how such knowledge made them more careful in sexual situations and by implication sexually healthier.

> It has allowed me to be safer in sexual situations. (AS, 18, Female)

> Because I know about what diseases etc. there are, I am more careful about it and who I have relationships with. (AS, 18, Female)

> Knowing about HIV/AIDS has made me think twice whenever I've wanted to have sex with a girlfriend. (AS, 19, Male)

> I know enough about sex and the consequences of it to avoid it unless I really love the woman it would be with, and plan to be with her for a long time. (AS, 18, Male)

Knowledge served as a warning that sexual activity was not always 'safe' and as a reminder to, as one young man puts it, 'think twice' before engaging in it. These responses are those sex educators might hope for from students who had absorbed programme messages about sexual risk and safety. However these comments also imply that safer sex messages may not have achieved quite the desired effect. While 'thinking twice' about the possible negative consequences of sexual activity, is a more encouraging safer sex behaviour than not thinking at all, acting upon these thoughts is another matter. Being 'more careful about who I have relationships with' and 'avoiding it unless I really love the woman' are not reliable disease prevention strategies. Moore and Rosenthal (1998) explain how young people in their research justified their non-use of condoms in relationships with the fact that they loved and trusted their partner and therefore had nothing to fear from them. While this may work within monogamous long-term relationships with uninfected partners, Moore and Rosenthal (1998) argue that this situation is likely to be 'idealised' and subsequently a 'flawed strategy for disease protection when multiple unprotected partnering occurs' (p. 238). Similarly, Waldby, Kippax and Crawford (1993, p. 30) described how young men in their study believed they could identify the 'sort of woman' who was likely to have an STI and based their condom use in relationships on a notion of 'clean' and 'unclean' women. When the young woman above states her sexual knowledge has made her more careful about *who* she has relationships with she evokes a similar notion of being able to determine a 'safe' or 'unsafe' sexual partner. These comments highlight the way participants read the knowledge they receive about safer sex through social discourses of 'love', 'sexual safety' and 'relationships'. Without insight into these conceptualisations of safety, official discourses of sexual knowledge may not reach young people in the ways intended.

Knowledge 'in' practice: negative consequences

While some young people conceptualised knowledge about 'sexual danger' as having a positive influence on their relationships, others described this as instilling fear. This reaction was most prevalent in young women's responses where knowing what could go wrong, made some feel anxious and apprehensive about sexual activity. The word 'scared' appears in each of the following extracts, conveying young women's fear of being 'used for sex', the possible pain of intercourse and the prospect of becoming pregnant.

> Because I am scared that my future boyfriends will use me for sex.
>
> (AS, 18, Female)

> I guess I'm too scared to have sex with my boyfriends, I hear it hurts the first time.
>
> (AS, 17, Female)

...I am scared of having a sexual relationship right now (pregnancy, etc) so I try to keep away from guys especially if I know they want sex.

(AS, 17, Female)

These responses draw on what Fine coins discourses of sexuality as *victimisation* and *violence*. Such discourses link sexuality with danger through their emphasis on sexually transmissible infections, unplanned pregnancy and various forms of sexual coercion (Fine, 1988). Within them young women are positioned as the victims of male desire as seen in the young woman's fear above that 'my future boyfriends will use me' and the suggestion that there is a need to 'keep away from guys' who want sex. In these narratives young women are sub-ject to an active male who will do things to them that they do not want. The discourse of sexuality as violence echoes in the other young woman's concern that sex will 'hurt' the first time, implying that the act of sexual intercourse is characterised by pain rather than pleasure. Fine argues that the culminate effect of these discourses is to incite 'terror' in young people in order to curb their sexual expression (Fine, 1988, p. 33). Clearly this form of knowledge has succeeded in inserting fear into the minds of these young women about the possible pleasures and dangers offered by sexual relationships.

Some knowledge about the dangers of sexual activity appeared to be gleaned not from knowledge imparted by others, but from first hand experiences. Two young women reported how knowledge gained through personal experiences they did not invite or want, had negative repercussions for their future sexual relationships. Their responses explained how knowledge of unwanted sexual attention and touch made them 'hate' sexual contact and be cautious in their relationship conduct.

I have been molested when I was young and never talked to anyone about it. I hate being touched by any guy, especially this person at work.

(AS, 17, Female)

From past experiences of sexual harassment, molestation, a married man having feelings, friends (boys) having feelings...this is why I can never be in a relationship and why I am slow in relationships. Past experiences, I suppose.

(NAS, 18, Female).

While hearing about sexual danger could serve a positive and pro-tective purpose for some young people, these responses suggest know-ledge gained through negative first hand experiences may impede future positive encounters for these young women. No young men referred to the experience of sexual abuse and its negative impact on relationships. This finding reflects the reality that young men are statistically much less likely to experience such abuse. For example, in Dickson et al.'s (1998) study of 13–21 year old New Zealanders only 0.2 per cent of young men in the study compared to seven per cent of young women, reported their first sexual experience was 'forced'. These patterns are further corroborated by responses to another survey question exploring young people's experience of unwanted 'sexual type touching'. Forty-one per cent of young women, compared to 16 per cent of young men reported experiencing such touching.

Conclusion

This chapter has revealed that relationships feature prominently in young people's lives and that they are an important context for the playing out of sexual subjectivities. Providing insight into how some young people describe their experience of them serves to contribute to our understandings of young people as sexual agents. Given the paucity of information about 17–19 year olds relationships in New Zealand, a more comprehensive picture of their experiences may prove useful for informing programme design and delivery around sexual issues. For instance, understanding young people's own con-ceptualisations of what constitutes a relationship provides clues to their reasons for entering into, and investment in them. Such informa-tion is vital for shaping health promotion messages in ways which reflect young peoples' lived experiences and are subsequently more likely to capture their attention to positive ends. Data on the fre-quency, length and character of these relationships may also help to centre these youthful conceptualisations in ways which offer alter-native discourses to representations of these partnerings as always 'short and sweet'.

In recognising young people as sexual subjects who experience desire and seek pleasure it has been important to document what young people found pleasurable about sexual relationships and how desire was expressed within them. This record serves to counteract discourses which deny sexual pleasure and desire as positive forces in young people's lives and as something which they should contain or deny until marriage (or later life). Documenting these experiences acknowledges desire and pleasure as lived realities of young people's relationships and serves to legitimate them as such. Charting sexual practices which young people find pleasurable highlights the kinds of behaviours they engage in and offers valuable contextual information for those who wish to promote safer sex.

Young people's relationships are not only the site in which pleasure and desire are played out but also a space in which gendered power operates. Drawing attention to how issues of contention are negotiated by partners, reveals much about the micro-politics of relationships. While couples in this study portrayed the operation of power in their relationships as equal, this was not always the case. Instead, power could also be seen to be mediated by young women and used coercively by young men. The complexity with which this power was manifested in relationships suggests that calls for safer sexual practices will only be effective if they take its fluidity and multifarious operation into consideration.

If we are to gain a better understanding of why young people do not always put the knowledge they acquire from sexuality education into practice then, enquiring how they perceive the relationship between knowledge and practice is imperative. Questionnaire participants saw that sexual knowledge could have positive implications for their sexual practice, such as alerting them to possible dangers and enabling them to have more sexually fulfilling relationships. However, not all sexual knowledge especially that which is constituted within discourses of victimisation and violence, was perceived to have positive effects. This sort of knowledge which as Fine (1988) puts it, 'weaves though the curricula, classrooms, and halls of public high schools' (p. 33) incited trepidation in young women particularly, causing them to report being fearful and guarded in relationships. These discourses offer young women and men limited and constraining ways of being sexual which are ultimately not fruitful to either. Such data once again supports the need for a discourse of sexuality which recognises young people as sexual subjects who have the potential to experience sexual desire and pleasure as a positive force in their lives. The next chapter begins the task of building such a discourse around what I have coined 'erotics'.

7
Constituting a Discourse of Erotics in Sexuality Education

This chapter presents a case for the inclusion of a discourse of erotics in sexuality education. I begin by delineating the concept of a discourse of erotics and the way this builds on Fine's (1988) seminal work around the missing discourse of desire. While it has been established that positive and empowering discourses of desire and pleasure are 'missing' from sex education, arguments about why their inclusion is important are less well articulated (Gagnon and Simon, 1973). To redress this situation the second part of the chapter briefly recapitulates the current research findings which support the incorporation of a discourse of erotics within sexuality programmes. This discussion proposes that a discourse of erotics may have gendered benefits for young women and men's experience and expression of sexuality. The final section considers the implications of a discourse of erotics for particular sectors of the youth population such as lesbian, gay, bisexual, transgendered, and intersex youth (LGBTI), those from different cultural and religious backgrounds as well as young people with disabilities. My aim is to begin a conversation about the possibilities of opening up discursive (and literal) space in schools for recognition of young people's sexuality as a legitimate and positive force, and not simply an element of identity which is denied or viewed negatively. This project takes cognisance of Britzman's (1998) observation that '....educators have yet to take seriously the centrality of sexuality in the making of a life and in the having of ideas' (p. 70). I argue that taking sexuality seriously is about promoting a sex-positive (as opposed to a predominately sex-negative) approach to sexuality education so as to effectively foster young people's sexual health and well-being.

Conceptualising a discourse of erotics

The idea of including a discourse of erotics in sexuality education extends Fine's (1988) identification of a 'missing discourse of desire' which positively acknowledges and incorporates young women's sexual desire at school. She explains that within official school culture, female adolescent sexuality is defined exclusively in terms of disease, victimisation and morality without positive acknowledgement of young women's subjective feelings of sexual desire and pleasure. Fine proposes that 'a genuine discourse of desire would invite adolescents to explore what feels good and bad, desirable and undesirable, grounded in experience, needs, and limits...would enable an analysis of the dialectics of victimisation and pleasure, and would pose female adolescents as subjects of sexuality' (p. 33). The concept of a discourse of erotics incorporates these elements while extending them in several distinct ways.

Like Fine, I assert that it is imperative for schools to enable students to 'explore what feels good and bad, desirable and undesirable' and that young women are positioned as active sexual subjects (Fine, 1988, p. 33). To achieve this objective may mean acknowledging young people as sexual subjects in positive ways and *legitimating* this identity rather than ignoring or punishing it. If schools could sanction a positive sexual identity in which young people are acknowledged as responsible sexual agents more opportunities for students to adopt this perception of sexual self might result. As schools currently operate, the only sexual identities young people are offered are negatively conceived in terms of sexual irresponsibility, promiscuity or as the victims of unwanted sexual advances. To legitimate young people as sexual subjects in positive ways requires a sex-positive rather than exclusively sex-negative approach to sexual activity, and incorporation of sexual desire and pleasure as commonplace instead of problematic. Viewing young people positively as sexual subjects also requires recognition that they are sexually embodied that is, they are subjects comprised of sensual flesh.

A discourse of erotics might incorporate an explicit recognition that *all* young people, what ever their gender and sexual identity (i.e. transgender, intersex, female, male, lesbian, same-sex attracted, bisexual, (hetero)sexual or something else) have a right to be positively acknowledged as sexual subjects who may experience sexual desire and pleasure. This inclusion makes reference to sexual diversity politics and a burgeoning literature documenting the invisibility of young LGBTI within school cultures and curriculum content. Researchers have described how sexuality education has traditionally ignored young LGBTI's needs for knowledge and affirmation of their sexual identities to considerable detriment to some (Vincent and Ballard, 1997; Quinlivan, 1996; Epstein, 1994; Hillier, Dempsey, Harrison, Beale, Matthews, and Rosenthal, 1998). In those instances when same-sex attraction is mentioned this is typically in relation to gay men and invoked negatively by connecting the transmission of HIV with homosexuality. In order that a discourse of erotics does not contribute to heteronormalising techniques that render heterosexuality as 'normal' and 'homosexuality' as 'deviance', it would need to contain an explicit affirmation of sexual and gender diversity politics. Inclusion of queer theory insights concerning the fluidity of sexual and gender identities might enable more flexible and less rigidly defined conceptualisations of self.

A discourse of erotics may also be about enabling young people's legitimate exploration of what might constitute their knowledge/ practice of desire and pleasure and a recognition of their right to knowledge that may facilitate this. The term erotics is conventionally defined as 'of concerning, or arousing sexual desire or giving sexual pleasure' (Sinclair, 1995). For subjects in this research it also comprised details about how to instigate or manage interaction that involved or might lead to sexual activity (e.g. how to ask someone out or how to physically engage in sexual activity). The presence of a discourse of erotics would render it legitimate to provide knowledge about the body as related to sexual response and pleasure and the logistics of bodily engagement in sexual activity. To avoid this discourse contributing to a standardisation of particular sexual practices or promoting a pleasure imperative, sexual activity and sensuality should be conceived in their broadest sense. Information about the logistics of sexual activity not only legitimates sexual desire and pleasure, but it is also vital for practising safer sex when teaching how and why particular activities are high/low risk for sexually transmissible infections. Given that young people justified their need for these details in order to improve their own sexual experiences, this information may also enhance interpersonal relationships.

Including a discourse of erotics in sexuality education could also be about creating spaces in which young people's sexual desire and pleasure can be legitimated and positively integrated within official school culture. This does not mean that young people have to, or will necessarily seize upon these spaces, but that they are no longer denied them because they are 'missing' from sexuality education programmes. All discourses have regulatory effects and it is important that a discourse of erotics does not render the experience of desire and pleasure compulsory or constitute those who don't experience these things as somehow lacking. Such consequences would be counterproductive to the purpose of having a discourse which acknowledges young people as sexual agents.

Another issue around the conceptualisation of this discourse is that it will take diverse forms depending on contextual factors like policy governing the teaching of sexuality, teacher training and class composition (age, ethnicity etc.). This diversity is necessary if it is to perform the task of creating spaces in which all young people might legitimately and positively explore 'what feels good and bad, desirable and undesirable, grounded in experience, needs, and limits' (Fine, 1988, p. 83). It is therefore impossible to give details of this discourse, given that as Hollway describes discourses are 'an interrelated system of statements which cohere around common meanings and values [that] are a product of social factors, of power and practices, rather than an individual's set of ideas' (Hollway cited in Gavey, 1992, p. 327). From this perspective a discourse of erotics cannot be outlined in terms of, or collapsed into, a list of possible curricula content. Neither is it authored or controlled by any one individual as it intersects with power structures, processes and other discursive/material practices beyond them. What I have outlined is my conceptualisation of how this discourse might look in order that it might contribute to social/ sexual justice.

Findings which support the inclusion of a discourse of erotics in sexuality education

The current research signals a number of reasons why a discourse of erotics may increase the effectiveness of sexuality education programmes. In Chapter 3 I argued that the concept of a 'gap' between knowledge gleaned from sexuality education and young people's actual sexual practices, does not acknowledge young people's own conceptualisations of sexual knowledge. The logic underlying the conceptualisation of this 'gap' is that what is learned in sexuality education is deemed necessary for sexual practice, a notion which many young people did not share. These young women and men's narratives indicated that being sexually knowledgeable meant being knowledgeable about a discourse of erotics. This kind of knowledge was given precedence over more 'official' discourses about for example, safer sex and was identified by young people as missing from sexuality education. Including a discourse of erotics within sexuality education may not only acknowledge these young people's own conceptualisations of sexual knowledge, but bring its teachings closer to their own interests and concerns. Programmes that don't meet young people's needs and interests are likely to fail to engage their attention and effectively communicate messages that will be enacted (Aggleton and Campbell, 2000). Including a discourse of erotics may be a way of closing the so called knowledge/practice 'gap' by bringing the content of such programmes into closer alignment with young people's own requirements and concerns. On a practical level, this might involve reformulating messages about sexually transmissible infections (for example) in ways which acknowledge and affirm young people as sexual subjects who can experience sexual desire and pleasure.

Another finding which supports the inclusion of a discourse of erotics involves the constitution of participants' sexual subjectivities. Young people revealed a more complex and nuanced understanding of themselves as sexual subjects than traditional notions of feminine passivity and active male (hetero)sexuality allow. Young women's narratives suggested that a significant number understood themselves as desiring sexual subjects who might initiate and exert a measure of control over sexual activity in their relationships. Sexuality education programmes that do not reflect this understanding these young women have of themselves, will be less likely to address them in meaningful ways. Programmes in which a discourse of erotics is absent, automatically cast young women as the passive recipients of active male desires. This positioning results from the gendered nature of (hetero)sexuality which constitutes young women as possessing lower levels of sexual desire and as being able to experience sexual pleasure less easily than young men. Without providing a discourse which positively recognises young women as active and desiring sexual subjects who have the potential to experience pleasurable corporeal feelings, young women are officially refused this sense of sexual self.

A failure to acknowledge the complex nature of young women's sexual subjectivities has particular ramifications for their sexual health and well-being. Without exploring the possible benefits of sexual activity in terms of pleasurable sensation, sexuality education neglects to provide young women with a standard against which to make decisions about engaging (or not) in this activity (Bollerud et al., 1990; Warr, 2001). Knowing whether sexual activity is wanted can be guided by knowledge that it will feel (or could feel) pleasurable. Without a sense of this pleasurable entitlement young women may elapse into sexual activity without an active sense of wanting this, or feeling they have anything to gain from this experience. By not acknowledging young women as sexually desiring subjects and revealing the possibility sexual activity might have pleasurable corporeal outcomes, sexuality education fails to convey a sense of personal empowerment for them. As others have noted (Holland Ramazanoglu, Scott, Sharpe, and Thomson, 1991a; Lees, 1993; Stewart, 1999) this has important repercussions for young women's sense of being able to initiate safer sex in relationships.

The absence of a discourse of erotics means that young women are not only discursively constituted as passive, but the possibility that they might experience sexual activity as corporeally pleasurable is also eroded. With no acknowledgement that young women's bodies can experience pleasure during sexual activity and that this is a positive outcome, the possibility of this being their experience may be reduced. This outcome can be explained by the way that discursive constructions are thought to have 'real effects' when language is seen to constitute meaning and possibilities for practice (Weedon, 1987). From this perspective women's greater likelihood of experiencing unsatisfying sexual activity, is not explained by their bodies being in some way essentially less easily pleasured than young men's, but that language constitutes their experience of them as such. This reality is witnessed in numerous studies documenting young women's dissatisfaction and disillusionment with (hetero)sexual practices (Thompson 1990: Hillier, Harrison, and Bowditch, 1999; Thomson and Scott, 1991). Given that the experience of sexual pleasure can have physical and mental health benefits, any omission to convey this to young women may have negative effects for their sexual well-being.

Young men's (hetero)sexual desire and pleasure appear to be given more space in some sexuality education programmes. Their desire is insinuated in information about 'wet dreams' and 'erections' and framed in a heteronormative discourse of 'growing up' and becoming interested in 'the opposite sex'. These discourses around awakening male sexuality also have a regulatory effect though, in their prescriptions of 'normal' and 'expected' (hetero)sexuality. Allusions to male (hetero)sexual desire in the absence of equivalent references to young women, constitute young men's sexuality as predatory. This representation provides an expectation about how young men should express their sexuality which may not always equate with their own feelings or experience. Young men can subsequently experience pressure to act in ways that are expected rather than desired, and have their expressions of sexuality limited to those deemed appropriately masculine. Narratives from young men in this research revealed responses that did not always adhere to traditional male subjectivities and instead constituted a 'softer' masculinity in which their enjoyment of romance and need for love were communicated. A discourse of erotics that embraces an understanding of sexuality and gender as more flexible and fluid than currently conceived, would provide opportunities for young men to express their sexuality in a broader range of ways. In this way, spaces for young men to move beyond the constraints of hegemonic masculinities may be maximised.

As indicated above, the presence of a discourse of erotics could mean recognition that young people are legitimate sexual subjects and that this encompasses a sense of their sexual *embodiment*. In Chapter 5 I argued that the sensual body is often omitted in sexuality education programmes where emphasis is placed on the reproductive and mechanistic functions of corporeality. To invite a thorough exploration of pleasure and desire as envisaged by the inclusion of a discourse of erotics, requires recognising the sensuality of the flesh and identification of actions which may produce or enable this. Findings revealed that young people who were sexually embodied indicated the most potential for acting upon safer sex messages promoted by sexuality education programmes. For young women sexual embodiment implied a sense of agency which is integral for exercising control over their sexual well-being particularly in terms of negotiating sexual activity with partners. Young men who are sexually embodied may also be more likely to translate knowledge acquired from sexuality education into practice because this state appeared to be achieved through emotional attachment to a partner. Young men's investment in relationships may be conducive to reinforcing the need to protect their partners and themselves from unwanted outcomes of sexual activity. While these instrumental aims of sexuality education might be served by a discourse of erotics which constitutes a sensual corporeality, this is also the basis for any comprehensive exploration of young people's desires and pleasures.

Chapter 6 revealed that power was gendered in young people's relationships and sometimes exercised coercively over young women or mediated by them in complex ways. Blanc (2001) highlights how the operation of power in relationships is closely linked to sexual health in terms of women's ability to negotiate condom use with partners. As a result of the way discourses are constituted within particular configurations of power they are never neutral. As proposed here, a discourse of erotics could contain an understanding of the way in which desire and pleasure are played out at the intersection of gendered power relations. It emerges from an explicit recognition that young people's sexual subjectivities are gender differentiated and the way they are enacted in (hetero)sexual relationships is often not equitable. A discourse of erotics could contain an assumption that this situation requires remedying and endeavour to foster relationships in which each partner (regardless of their gender) is empowered to make decisions which benefit each person's sexual well-being.

An argument for promoting gender equitable sexual relations emerges from findings about young men's preferred sources of sexual knowledge. These data revealed that almost three quarters of young men in the sample had consulted pornographic magazines and rated this information 'very useful'.[1] In the absence of a discourse of erotics in sexuality education young men may seek such information in other contexts and find it most overtly available within pornography. Anti-pornography feminists have written prolifically about the effects of pornography and the way it can exploit and objectify women (Coward, 1987). Mainstream pornography is a key mechanism for the constitution of hegemonic masculinities and this construction is seen as deleterious for sexual equality. As Jackson and Scott (1996) have argued, pornography

....helps to circulate and perpetuate particular versions of these narratives such as the mythology of women as sexually available, deriving pleasure from being dominated and possessed and a model of masculinity validated through sexual mastery over women. A man does not rape as a direct reaction to a pornographic stimulus; rather pornography contributes to the cultural construction of a particular form of masculinity and sexual desire which make rape possible.   (p. 23)

Pornography as a source of information about (hetero)sexual desire and pleasure can reproduce oppressive male sexual subjectivities that sustain inequitable sexual politics (Dworkin, 1981). It also offers young men a limited range of conventional ways of being masculine subjects which deny them other forms of expression of desire and experiences of pleasure. By neglecting to include a discourse of erotics that incorporates (hetero)sexual desire and pleasure in a positive, empowering and gender equitable manner, sexuality education offers no alternative comparable discourses with which to contest mainstream pornography.

Without these, many (hetero)sexual young men may only have access to discourses of pornography that not only limit their own experience of sexual desire and pleasure, but also those of their partner.

Implications of a discourse of erotics for particular sectors of the youth population

Young people with disabilities

In this section I consider what a discourse of erotics might mean for young people with disabilities. Where positively recognising young people as sexual subjects is fundamental to a non-disabled youth population, it is imperative for young people whose bodies don't conform to dominant perceptions of able-bodiedness. For these young people acknowledgement as legitimate sexual subjects is even more challenging because their disabilities are seen to render them dualistically as either asexual or sexually inadequate. This 'myth of asexuality' as Milligan and Neufeldt (2001) coin it, is connected to two deep held assumptions. The first is that for people with physical disabilities actual or presumed 'sexual dysfunction' prevents or lessens the need for sexual gratification and expression. Secondly, as Milligan and Neufeldt (2001) explain, 'although their sexual function is typically intact, individuals with intellectual disabilities and/or psychiatric disorders are thought to have limited social judgement and therefore, lack the capacity to engage in responsible sexual relationships' (p. 92). As a consequence of this thinking these people are perceived as infantile and like children, are presumed not to know enough to make informed decisions about sexual matters. When people with disabilities *are* acknowledged as sexual beings, this is often in terms of sexual deviance for example, inappropriate sexual display or masturbation (Shakespeare, Gillespie-Sells, and Davies, 1996, p. 10). Such representations assign people with disabilities as inferior to a non-disabled population whose sexuality is thought more sophisticated and less offensive.

A discourse of erotics which positively acknowledges young people with disabilities as sexual subjects would draw on the disability movements' call for societal recognition of their full personhood. Disability politics has highlighted the way in which disability is socially constructed and not biologically determined so that people with so called 'impairment', are disabled by society and not their bodies. The problem of disability is then shifted away from the constructed 'deficits' of disabled people themselves and located firmly in society's inability to accept and accommodate its differently able citizens. Within this social model of disability:

> The main 'problem' of spinal injury is not a failure to work normally, but a failure to gain access to buildings if one uses a wheelchair. The difficulty of deafness is not inability to hear, but the failure of society to provide Sign Language interpretation and to recognize deaf people as a cultural minority (Gregory and Hartley, 1991 cited in Shakespeare et al., 1996, p. 2).

Through this treatment society denies people with disabilities the status of 'full personhood' which is automatically accorded to other non-disabled adults where sexual expression is perceived as normal and natural. The connection between sexuality and full personhood is revealed by Shakespeare et al. (1996) when they explain that within 'modern Western societies, sexual agency (that is, potential or actual independent sexual activity) is considered the essential element of full adult personhood' (pp. 9–10). When people with disabilities are not recognised as sexual they are denied the same rights and status as those who are non-disabled. A discourse of erotics which renders all young people as sexual agents has significance beyond being recognised as sexual for those with disabilities, it is about being counted as a fully operational human being.

Benjamin Seaman declares that access to sexual pleasure is 'the real accessibility issue' for people with disabilities. Sexual politics is often left off the welfare agenda however in favour of what the disability movement has generally identified as more fundamental issues such as housing, education, transport and employment (Finger, 1992 cited in Davies, 2000). This hierarchialisation of human rights is reflected in wider society, where sexuality is perceived as separate from and unimportant to survival in daily life. Tackling issues around sexuality and disability would mean examining 'the real accessibility issue' because of their status on the bottom rung of the hierarchy. A discourse of erotics which asserts young people's right to sexual pleasure and their expression of desire in positive ways would recognise people with disabilities as 'full human subjects' and could have other benefits. Tepper (2000) argues, that for people with disabilities sexual pleasure

> ...can add a sense of connectedness to the world or to each other. It can heal a sense of emotional isolation so many of us feel even though we are socially integrated. It can help build our immunity against media messages, that can make us feel as if we don't deserve pleasure. (p. 288)

Within a discourse of erotics sources and expressions of sexual pleasure could be conceptualised beyond the coital imperative as a means of widening the possibilities of what might be constituted as sexually pleasurable (McPhillips, Braun, and Gavey, 2001). While the coital imperative regulates and limits the sexual expression of the non-disabled population it is particularly oppressive to people with disabilities 'who cannot operate according to "fucking ideology", because of difficulty with positioning, erectile dysfunction etc' (Shakespeare et al., 1996, p. 97). This point is poignantly made by one woman in Shakespeare et al.'s (1996) research who explained that, 'My dream is to open a magazine and think "I could try that", instead of saying, "You must be joking – I can't even look up at the chandelier, let alone swing from it!" Information is power and disabled people still don't have enough of it' (p. 15).

Another component of a discourse of erotics could be young people's right to knowledge about the body as it relates to desire and pleasure. Research indicates that people with disabilities receive less sex education and are subsequently less knowledgeable about sexual matters than a non-disabled comparison group (McCabe, 1999). In addition, where as people from the non-disabled population were most likely to receive sex education from parents, friends and other sources, people with a disability, were most likely to only receive their sex education from 'other' sources (McCabe, 1999, p. 167). The reasons people with disabilities are often excluded from sexuality education are because (as indicated above) they are perceived as asexual or are infantilised. As Milligan and Neufeldt (2001) suggests and McCabe's (1999) study reveals, in their efforts to protect a child with disability from 'future rejection, vulnerability to sexual abuse, or unwanted pregnancy, well-meaning parents and professionals may ignore the topic of sex' (Milligan and Neufeldt, 2001). The right to knowledge about sexual pleasure and desire as conceptualised by a discourse of erotics could mean information tailored to young people's specific needs and situations as dictated by their disability. This discourse could acknowledge that disability is diverse with a range of physiological and emotional effects which configure bodies and people in a multitude of ways. As Davies (2000) explains:

We may need for example to learn about specific positions which would enable us to give and receive sexual pleasure, ways of managing pain and spasm, how to deal with the physical adaptations which enable us to function in the world but which may become obstacles in sexual relationships.   (p. 189)

When people with disabilities were asked who should impart this information, those in Shakespeare et al.'s research revealed a preference for 'other people with disabilities'. A discourse of erotics which conceptualises sexual activity and sensuality in their broadest sense could have specific significance for people with disabilities. Instead of feeling they are lacking or missing out on sexual pleasure gained from activities outside of their bodily repertoire, a discourse of erotics has the potential to position people with disabilities in more empowering ways. This potential arises because they are not confined by conventional 'fucking ideology' and embody a corporeality that renders greater opportunities for creative forms of sexual pleasure. Within such a discourse, people with disabilities are constituted as sexual subjects with the greatest potential for the attainment of sexual pleasure which disrupts the coital imperative. The bodies of people with disabilities encompass the potential for a new kind of sexual revolution.

A discourse of erotics which embraces the complexities of sexual diversity politics may redress a common misconception that people with disabilities are all heterosexual. While being seen as a sexual subject is more difficult for people with disabilities the idea that they might be gay, lesbian or bisexual is often never considered by their carers. Such heterosexism, to which non-disabled gays and lesbians are also subject has particular ramifications for people with disabilities. Those who live in community homes whose rules or caregivers are unsupportive of their gay, lesbian or bisexual identity may find few opportunities to express their sexuality. Similarly, as Shakespeare et al. (1996) explain, 'if heterosexual disabled people find it difficult to find personal assistants who will help them engage in sexual activity, the difficulties experienced by lesbians and gay men are significantly greater' (p. 167). Disabled lesbians and gay men may find themselves stranded when their carer discovers their sexuality or live in constant fear about being found out (Shakespeare et al., 1996). A discourse of erotics which draws attention to people with disabilities as diverse rather than uniform sexual subjects may counter misconceptions that when young people with disabilities are deemed sexual at all, they are heterosexual.

Queer youth

What a discourse of erotics might mean for queer youth is acknowledgement and legitimation of their identities which have generally remained hidden or been silenced in schooling contexts. Schools have traditionally not been safe spaces for these young people who often feel alienated by educational structures and processes which promote and naturalise heterosexuality. As a consequence the queer subject is typically missing from the formal curriculum and often only referred to with derision in informal school cultures (e.g. playgrounds, staff-rooms). When this group of young people are recognised as sexual sub-jects they are typically constituted as 'other' to the presumed default of heterosexuality. Queer identities are not seen as a usual expression of sexual diversity but rather an alternative to heterosexuality. Researchers' attempts to redress the invisibility of queer youth at school have some-times (unwittingly) contributed to a negative constitution of them. This consequence has occurred by drawing attention to the difficulties these young people can face such as their experience of victimisation and early death. When these factors are emphasised over and above the strength and resiliency of this youth population their effect is a seemingly disempowered subject. This is not a sense of self from which queer youth might draw strength and a perception which may con-tribute to their victimisation. A discourse of erotics which enables queer students to be visible in schools in a positive way that is not con-stituted negatively as 'other', could signal a significant shift in the recognition of these young people.

A discourse of erotics which recognises the fluidity of sexual and gender identities might offer queer youth freedom from using stigma-tised labels which they often reject. Savin-Williams (2001) notes that young people are often unlikely or unwilling to classify themselves as lesbian, bisexual or gay. This disinclination can occur because 'even among youth with same-sex attractions many do not believe that they fit the definition of that label, dislike the political or sexual asso-ciations with the label, or feel that the terms are too simplistic or reductionistic to describe their sexuality' (p. 10). Examining the devel-opment of same-sex attracted identity in young women, Diamond and Savin-Williams (2000) found that these ran 'counter to the con-ventional view of sexual orientation as a stable, early appearing trait' (p. 298). Instead, exclusive same-sex attractions were an exception rather than the norm among those in the study with some vacillation between same-sex and other-sex attractions across time. Recognising that young people may have a sense of fluidity about their sexual attraction and may not subscribe to gay or lesbian identity has impor-tant ramifications for how sexuality education's messages are delivered. Safer sex messages that name and target gay and lesbian youth may not reach all of their desired audience if young people do not recognise themselves by these labels. A young man who has a regular girlfriend yet engages in casual sex with other men in clubs may when asked, identify as heterosexual. Health promotion messages concerning the use of condoms directed at gay men are unlikely to have resonance for this young man who views himself as heterosexual. A discourse of erotics which recognises the complexity and fluidity of homosexual (and heterosexual) identities could offer young people more flexible and perhaps less stigmatised perceptions of sexual self.

An issue related to identity arises here and involves the transgendered and intersex communities. While the fluidity of sexual identities might be a foreign concept for the general population, it is less so than the idea that gender need not be expressed in terms of dualisms. The novelty of this conceptualisation was reinforced for me recently when designing a questionnaire for use in secondary schools. In the item asking participants to label their gender, I included the options of 'both/neither' and 'something else' after the usual 'female', 'male'. In every class in which I personally distributed the questionnaire students (and sometimes teachers) would ask 'how can you be both male and female or neither'? These questions sparked discussion about people who are intersex and other members of the transgendered community who may identify their gender in unconventional ways. From these discussions young people's confusion in separating out issues of gender and sexuality were apparent. This perplexity is not surprising given that gender and sexuality are intricately connected, but many had difficulty in recognising that people who are transgendered were most likely to describe themselves as heterosexual. Students typically conflated transgendered or intersex with being gay a misrecognition that sexuality education must address if young people who are not members of these communities are to have a more comprehensive understanding of them. A discourse of erotics which acknowledges those whose gender transcends traditional categories of male and female could help to legitimate these identities. Such a discourse could also enable a conceptualisation and expression of gender that was less restricted by traditional gender dualisms.

While pleasure and desire have been missing from sexuality education for young heterosexuals, the denial of queer youth's very presence at school has meant these topics haven't even constituted a whisper. In terms of sexual safety an important omission has been safer sex information for same-sex sexual activity. In a letter to SEICUS, 17 year old Liz Conley (2003) describes what typically occurs in sexuality education classes where there is an assumption that all class members are heterosexual.

Maybe I was just looking for something that wasn't there, but the overwhelming emphasis of the whole presentation was on hetero-sexual sex. Even anal and oral sex were only mentioned with regard to male/female relations, and every mention of male condoms other than their application was addressed to females....All I want is straight-forward information without a stigma attached to it and acknow-ledgement of sexualities outside of heterosexuality. There's a whole population out there – 10 per cent of us, in fact – who could be educated about how to avoid STDs. Is that so much to ask?   (pp. 4–5)

A discourse of erotics could legitimate this right to information about sex and sexuality and the inclusion of pleasure and desire as relevant to same-sex attracted youth. Including such material should not necessitate pref-acing it with for example, 'if you are a homosexual you will need to use a condom for anal sex'. As discussed above, such an approach may be counter-productive for those who choose not to recognise themselves by this label. Framing sexual information for queer youth in this way may also reinforce a sense of stigmatised otherness especially when such com-ments are tacked on to discussions about vagina/penis intercourse. There are other issues which might be included within a discourse of erotics per-taining to same-sex attracted youth that are vital for a positive experience of sexuality and in maintaining sexual health. One possibility here, is the pressures young gay men who are not sexually experienced may encounter to engage in receptive anal sex.

As it emerges from this research a discourse of erotics is about social justice. To this end it involves recognising and valuing sexual and gender diversity (as well as gender equality) and claiming the right to positively experience sexual pleasure and desire for all young people. To achieve this objective it is imperative that seeking to address homo-phobia is integral to a discourse of erotics. Homophobia is described as the fear/hatred of same-sex attracted people or fear of being perceived as same-sex attracted (Eadie, 2004). The expression of such fear can occur at the level of individual attitudes and institutionally in the form of policies where heterosexual norms are the standard against which other ways of thinking and 'being' are assessed. Schools are sites in which homophobia is expressed on both these levels through abusive comments between individuals and heteronormative institutional practices which assume all students are parented by heterosexual parents. Such homophobic practices foster a culture that not only con-tributes to a negative sense of sexuality but an intolerance of sexual difference. A discourse of erotics should work to disrupt homophobia by positively embracing sexual (and gender) diversity by means that do not 'other' those that do not conform to conventional norms.

Youth from different cultural and religious backgrounds

This section considers the implications of constituting a discourse of erotics within sexuality education for young people from particular religious and cultural backgrounds. How to reconcile a diversity of cultural beliefs and religious values within sexuality education poses a considerable challenge to the delivery of this curriculum area (Reiss, 1997). The complexity of this task is attributable to the moral values contained within certain religious and cultural perspectives which can conflict with the very idea of sexuality education in schools, let alone programme content. McKay (1997) characterises these cultural values and religious morals in terms of two divergent sets of sexual ideologies known as 'restrictive' and 'permissive'. Those from religious and cultural backgrounds that adhere to a restrictive ideology, view sexuality from a negative standpoint and sexual behaviour as requiring stringent legal, social and moral controls. Only a narrow set of sexual behaviours are deemed morally acceptable so that sexual activity beyond procreative intercourse within a monogamous marital relationship is considered wrong (McKay, 1997). Groups who subscribe to 'permissive' sexual ideology view sexual behaviour as 'a natural and pleasurable aspect of life which is meant to be enjoyed' (McKay, 1997, p. 287) and contributes to personal fulfilment and health. Provided consent is mutual, and conduct sexually responsible, 'permissive' sexual ideology endorses non-procreative sex and considers same-sex attraction and premarital sexual activity morally acceptable. It is arguable whether these sets of ideas are 'ideological' or 'discursive' and pertinent that their conceptualisation as 'restrictive' and 'permissive' implies a certain moral perspective. However, these categorisations of beliefs about sex and sexuality capture two distinct sets of meanings which coincide and conflict with the interests of particular ethnic, religious and sexual minorities.

The opposing nature of these sets of sexual discourses has emerged in public debates about 'values' and 'morals' in sexuality education. Concern around the morals and values that sexuality education embodies has received more sustained attention in countries like Britain where faith based communities are more diverse and have greater volume (Thomson, 1997). In New Zealand, concern with a moral and values framework in sexuality education has been confined to the intermittent lobbying of a relatively small pool of religious groups. However, the ethnic composition of New Zealand's' population has become more diverse in recent years with people from variegated ethnic and faith based backgrounds (like Iran, Somalia, Tuvalu/Ellice Island, Society Island, Korea, China, Arab nations) entering the country in higher numbers (Statistics New Zealand, 2002). Many of these new arrivals have come from communities that prescribe a restrictive sexual ideology and this presents new challenges for sexuality education's delivery. Within McKay's conceptual framework a discourse of erotics fits more easily with so called 'permissive sexual ideologies'. Recognition of same-sex desire as 'natural', a sex-positive stance and constitution of young people as legitimate sexual agents makes a discourse of erotics abhorrent to some faith based and cultural communities. How then can the inclusion of this discourse be reconciled within sexuality education in this cultural climate?

Considering the alternative of the continuing absence of a discourse of erotics in sexuality education is one way of answering this question. Teachers seeking to avoid controversy have often circumvented topics of moral dispute (such as homosexuality) under the precept of appearing value neutral. The omission of these topics is anything but morally neutral however, communicating to students that these issues are unmentionable and therefore negatively conceived. Moral objectivity is an impossibility in sexuality education and can constitute an impediment to young people's basic right to health (e.g. when same-sex desire is bypassed and subsequently pertinent safer sex information denied). Emphasis on one moral or cultural perspective about sexuality is also a form of indoctrination that limits young people's agency and opportunities for sexual well-being.

We live in a plural society, reflected in the circulation of a multiplicity of discourses about sexuality. Within this social environment disagreement about sexual morals is inevitable, but such contention should not form the reason for sexuality education's paralysis. Thomson (2000) describes our current social climate as one of moral complexity but is careful to differentiate this from 'moral relativism'; she writes:

> Moral complexity is not the same thing as moral relativism. The danger of current developments in education policy in the area of spiritual, moral and social cultural development is that they made this equation and used it to justify a move towards a rhetoric of simplicity and certainty away from a reality of complexity and uncertainty. While such a strategy may be politically attractive it has little practical application in an increasingly diverse and complex world. Schools are part of this world and have to respond to this complexity.   (p. 268)

Responding to this complexity means allowing a diversity of discourses of sexuality within sexuality education, including a discourse of erotics which answers young people's call for information about sexual desire and pleasure. This request has not been confined to New Zealand with teenagers in British based research proposing Sex and Relationship Education (SRE) incorporate information on the logistics of sexual activity (see Measor et al., 2000). In continuing to deny these young people a discourse of erotics, sexuality education not only fails to meet their needs and interests but is unlikely to be successful in capturing them with messages intended to keep them safe. Blake and Katrak (2002) explain '...young people from all faiths and cultures have an entitlement to sex and relationships education (SRE) which supports them on their journey through childhood to adolescence and adulthood' (p. 1). The achievement of sexual well-being is contingent upon access to a multiplicity of discourses about sexuality which promote a positive view of sexuality and are underpinned by values of equality and respect. One young woman in Blake and Katrak's (2002) research emphasised the importance of receiving this entitlement to knowledge when she said:

> I have a faith and I trust my parents to talk to me about values. At school what I need in sex education is to understand about sex and relationships and understand what different people think.   (p. 1)

Including a discourse of erotics is not about supplanting other discourses of sex and sexuality at school. This possibility is unlikely given the way power intersects with discursive practices. Because discourses are multiply layered the introduction of 'new' discourses does not engender the eradication of others. Davies (1993) describes this process through the metaphor of palimpsest where new writings on a parchment were scribbled around or over old writings that were not completely removed. Davies explains this effect as where:

> One writing interrupts the other, momentarily overriding, intermingling with the other; the old writing influences the interpretation of the imposed new writing and the new influences the interpretation of the old. But both still stand, albeit partially erased and interrupted. New discourses do not simply replace the old as on a clean sheet. They generally interrupt one another, though they may also exist in parallel, remaining separate, undermining each other perhaps, but in an unexamined way.   (p. 11)

Those who believe a discourse of erotics will 'contaminate' youth and encourage their sexual promiscuity misunderstand the process of discursive constitution and underestimate young people's agency. Creating discursive space for a student subject that is positively and legitimately sexual does not mean all young people will take up this positioning. However, failing to present such as discourse reduces opportunities for those who will choose to be sexually active no matter what directives they receive, from benefiting from this subjectivity. As indicated earlier, it is a subject position from which sexual subjects are most likely to act in ways that will support their sexual well-being.

In her work with the Sex Education forum, Thomson (1997) describes developing a consensus statement with diverse faith based and cultural communities on the purpose and content of sex education. Valuable lessons were learned from this process about working with communities that hold divergent views about sex and sexuality. Thomson observed that when debates over values moved from the intangible arena of moral and religious philosophy to the practical realm of young people's lives, then greater consensus on sexual issues was achieved. She provides the example of 'acknowledging that people can be born into and identify with a religious and cultural tradition that condemns homosexuality, yet still be gay, and deserving of support and respect' (p. 267). As proposed here a discourse of erotics would be grounded in young people's sense of reality and the kinds of issues and interests they raise as important to their lives. Despite trying to adhere to the regulation of abstinence, for some young people the need to express their desire and seek sexual pleasure means this is an impossible expectation. A discourse of erotics might incorporate this reality and address young people's sexuality in a positive way. If these young people's sexual well-being is to be supported then this discourse can no longer be missing from sexuality education.

A final note

The effectiveness of a discourse of erotics in sexuality education will depend on a multitude of factors, ranging from the institutional support it receives from schools to the needs of students in the context it circulates. Another important factor in its success is the ability of those who teach about sex and sexuality to effectively deliver programme content containing this discourse. Young people are acutely attuned to the comfort and confidence of teachers in discussing sexuality issues and this skill can make the difference between students' engagement in, or dismissal of, the subject (Kehily, 2002; Munro and Ballard, 2004). Being non-judgemental, trustworthy, open and honest, respecting young people's rights to choices/decisions are qualities that students and teachers alike value in sexuality educators (Milton, Berne, Peppard, Patton, Hunt, and Wright, 2001). Training teachers in how to deliver this discourse in ways that are beneficial to young people will also determine the success of this discursive strategy. Understanding what a discourse of erotics is, the history of desire and pleasure as missing from sexuality education programmes as well as the potential benefits of such a discourse would be important components of this training. In addition to the qualities outlined above teachers' skill base may also include reflexivity about their own sexuality and sexual values and how these inform classroom pedagogy. This work is already underway in some of the teacher training institutions in New Zealand where educators see value in a discourse of erotics.

8
Closing Thoughts and Future Directions

To better understand the idea of the 'gap' phenomenon this study has explored young people's (hetero)sexual knowledge, subjectivities and practices. This investigation was motivated by a desire to improve sexuality education's capacity to enable young people to experience their sexuality in more positive and empowering ways. In this chapter I summarise some of the main research findings and assess their implications for better ways of conceptualising the design and delivery of sexuality education. It is not my intention to be prescriptive here, but rather to open spaces for other ways of thinking about the aims and possibilities offered by these programmes. The last section of this chapter considers some of the questions left unanswered by the research and points to possible areas for future exploration.

Main research findings

Findings concerning participants' own conceptualisations of their sexual knowledge, subjectivities and practices have resulted in a problematisation of the notion of the knowledge/practice 'gap' throughout this book. A main argument has been that this 'gap' is an adult conceived idea, the underlying logic of which young people often do not adhere to. This was evidenced in Chapter 3 when participants indicated that they did not deem sexual knowledge as a prerequisite for sexual practice. While the 'gap' equation suggests sexual knowledge is necessary to engage in sexual activity, many young people perceived sexual practice as a means of becoming sexually knowledgeable. Where the logic of the 'gap' equation collapses for participants is that it did not take account of their own conceptualisation of sexual knowledge which involved more than the official discourses about risk and disease typically offered by sexuality education. Participants revealed that the kind of sexual knowledge that they were most interested in could be characterised as 'erotics' and that this was missing from sexuality education programmes. Without acknowledgement of the information that participants want and find most compelling, the 'gap' phenomenon fails to reflect young people's realities. As a framework for thinking about the effectiveness of sexuality education it is not formulated in relation to young people's interests and concerns, yet professes to be concerned with their sexual health and well-being.

The research findings have also revealed the idea of the 'gap' phenomenon as problematic in the way that it denies young people a positive and active sexual subjectivity. By positing knowledge learned in sexuality education as necessary for sexual practice the 'gap' phenomenon constructs a youthful subject for whom this logic *should be* unproblematic. The idea that knowledge learned from sexuality education should be translated into practice without hiccups presumes a youthful subject who does what they are told. This construction denies young people as actively desiring sexual subjects who may be more motivated by physical or emotional concerns than the fear of sexually transmissible infections. Conversely, the very idea of a 'gap' phenomenon constitutes a subject that does not execute the application of knowledge into practice as a 'problem'. These young people may be seen as actively desiring subjects, but this construction is negative in that their desire is seen to lead them into danger. Both constructions portray young people as devoid of a positively construed sexual agency which enables their own desires to be met without negative consequences for themselves or others.

Despite this constitution of young people's sexual subjectivity within the idea of a 'gap' phenomenon findings in Chapter 4 reveal many participants viewed themselves as social actors who exercised control (at least at a perceptual level) over their relationships and sexual activity. This sense of agency was also revealed when their talk about sexual subjectivity drew on dominant discourses of (hetero)sexuality in ways which suggested their engagement with, rather than their adoption of such meanings. One particularly poignant instance, was when Anna reworked the meaning of 'slut' so that her own behaviour and sense of self fell outside of it. In these ways participants simultaneously accommodated and resisted dominant discourses of (hetero)sexuality rendering themselves not only sexual subjects but active meaning makers. Findings concerning the micro-politics of sexual relationships in Chapter 6 also indicate that young people have an active role to play in how knowledge gained from sexuality education is put into practice. These findings imply that the way young people are constituted by the idea of a 'gap' phenomenon, does not capture the active sense of sexual self many of them communicated.

As a consequence of these findings I have questioned the utility of framing concerns about sexuality education's effectiveness in terms of a knowledge/practice 'gap'. Formulating the problem of sexuality education in this way is indicative of a lack of understanding about how young people view themselves as sexual subjects and how they conceptualise sexual knowledge. For the instrumental aims of sexuality education to be achieved (i.e. sexually healthy, responsible young people who are able to experience their sexuality positively) schools need to carve spaces for the existence of this kind of sexual subject. If we see this conceptualisation of the 'gap' phenomenon as reflective of how young people are constituted as subjects at school, then such a sexual subject position does not exist. At school young people can only *legitimately* be non-sexual subjects who heed messages to delay sex before marriage and are academically successful. Conversely, they can be problematic sexual subjects whose sexual activity is seen to interfere with their academic pursuits and life chances. Re-conceptualising our ideas about sexuality education's effectiveness in ways that offer students a legitimate and positively construed sexual subjectivity might go some way in transcending the knowledge/practice stumbling block.

Another argument made in this book is the need to 'enflesh' our understanding of the 'gap' phenomenon by analysing how young people describe embodied sexual experience. In Chapter 5 this exploration revealed the gendered nature of sexual embodiment which I conceptualise in terms of a series of embodied states (embodiment, dysembodiment and disembodiment). The state of sexual dysembodiment represents an extension of existing embodiment theory and indicates how dominant discourses of masculinity and femininity effect young people's experience of the sexual body. This politicised theory of sexual embodiment reveals how each embodied state has particular consequences for young women and men's experience of sexual pleasure and exercise of power in sexual relationships. For instance, I propose the corporeal sensual detachment displayed by young women who were seen to be disembodied may mean they are less aware of what they want or are doing in sexual contexts. This state could affect their ability to assent to sexual activity or introduce the issue of condom use. These findings suggest that sexual embodiment is central not only to young people's sexual subjectivities but to thinking through the 'gap' phenomenon.

Data about the length, frequency and character of young people's relationships reveals that they can invest considerable time and emotional energy in these romances. This finding conflicts with the popular belief that young people's relationships are typically not serious and instead 'short and sweet'. It also highlights the importance of these relationships as a site where young people can be seen as sexual agents and for the playing out of sexual subjectivities. Documenting the things participants described as pleasurable about relationships serves as a means of developing an empirical basis upon which to positively acknowledge young people as sexual actors. The fact that young people named both physical contact and emotional aspects of relationships as pleasurable suggests it is often more than desire which motivates intimate involvement. Such information is invaluable for constituting a discourse of erotics within sexuality education based on young people's own experiences in order to draw these programmes into closer alignment with their needs and interests.

A salient finding about young people's relationships is recognition of how young people perceive power to operate within them. To conceptualise the types of power young people described, a theoretical framework of 'equal', 'mediated' and 'coercive' power was devised. While all of the couples suggested that power was shared equally in their relationship, their talk provided examples of a more coercive form at work. Young women's narratives insisted however that they exercised agency even in instances where they appeared to prioritise the pleasures and needs of their partners at the expense of their own. In these instances young women seemed to actively subject themselves to this power, viewing this as a form of control because of the active decision it involved. What was evident here was the complexity with which power manifested in their relationships and the multifarious and fluid manner in which it operated. In terms of thinking through the 'gap' phenomenon these findings indicate the importance of recognising that any knowledge gained from sexuality education is applied in a context of shifting and complex gender relations.

Implications for sexuality education

Based on what young people have revealed about their understanding of their sexual knowledge, subjectivities and practices this book calls for the inclusion of a discourse of erotics in sexuality education. Such data highlights the importance of taking our lead from young people in any re-conceptualisation of the design and delivery of sexuality education. Without understanding young people's own sense of their sexual knowledge, subjectivities and practices we have scant hope of engaging their interests and making sexuality education relevant to their lives. By centring young people's sense of their sexual selves we are more likely to provide them with information that they deem helpful and which they will apply in practice.

Sexuality education is one of the few places young people receive officially sanctioned messages about sex and sexuality. It therefore carries an authority which is not easily rivalled by information gained from friends or other informal sources such as the media. However, this power may be undermined when sexuality education's messages deviate significantly from what young people already know about sex and sexuality. When young people receive the message from school that sexual activity is predominately about danger, guilt and risk while elsewhere it is promoted as involving fun, pleasure and power, sexuality education's warnings can appear didactic and boring. The challenge is creating space in schools for a discourse of erotics which pushes at the symbolic boundaries of public/private, teacher/pupil, proper/improper which govern contemporary schooling.

I would argue that incorporating an understanding of how young people view themselves as sexual subjects reveals the necessity of a shift in the motivating philosophies of sexuality education. This means re-evaluating the adult conceived view of sexuality education programmes as about avoiding sexually transmissible infections and other negative consequences of sexual activity. The study's findings suggest that in order for these warnings to be heeded, a first step might be their reconfiguration within a discourse of erotics. A strategy schools might undertake is the reshaping of the purpose of sexuality education so that it is concerned more with the capacity to enable young people to experience their sexuality in positive and empowering ways. This doesn't mean instigating a pendulum swing from messages that render 'sex as bad' to 'sex as good'. Young people are likely to be sceptical of this approach given the revolutionary change it would represent from the kinds of things they have already learned about sexuality from school. Instead, it would involve rethinking the content and delivery of sexuality education in terms of positive experiences of sexuality for young people that are not confined to being disease-free.

If sexuality education is to accommodate young people's sense of themselves as sexual people then their sexuality needs to be legitimated at school and ultimately by society at large. We would therefore need to review our current treatment of young people's sexuality as a problem and distraction to their studies (Nash, 2001). Positioning students as positively construed sexual subjects opens space for them to take up this subjectivity. It legitimates their sexual selves while providing opportunities for them to be the kind of responsible, respectful and caring citizens schools endeavour to produce. Achieving this will require more than viewing young people in a new light, it also entails a more positive view of the nature of sexuality. In countries like Britain, America, Australia and New Zealand sexuality is generally an uncomfortable subject, about which a sense of secrecy and controversy pervades. When overt displays of sexuality occur in these cultural contexts they are often seen as something vulgar and/or perverse. Public conceptualisations of sexuality differ markedly in countries like the Netherlands which are touted as enviable examples of low rates of teenage pregnancy and sexually transmissible infections. The success of countries like the Netherlands is attributed to good social infrastructure enabling easy access to contraception and condoms, as well as a particular public attitude towards teenage sexuality (Innocenti, 2001).

> It is the openness and the acceptance that young people will have intimate sexual relationships without being married and that these relationships are natural and contribute to maturing into a sexually healthy adult. It is the refusal to brand the expression of sexuality as deviant behaviour or to cast it solely in a negative light. It is the determination to present sexual expression as a balance – a normal part of growing up *and* a responsibility to protect oneself and others. It is the respect these societies have for adolescents, valuing them as much for who they are as for the adults they will become. (Berne and Huberman, 1999, p. vi)

Transforming social perceptions of sexuality is beyond the capabilities of any sexuality education programme and instead requires a sea change in public attitudes. Schools might begin to ignite this change by creating an approach to young people's sexuality in their curricula and through their treatment of students that is sex-positive rather than sex-negative. This approach entails constructing sexuality and sexual activity as something that is positive rather than only a problem, and young people's experience of these things as a valued and accepted element of their identities.

Connected with this need for schools to position students as legitimately and positively sexual, is the importance of recognising the complexities inherent in young people's sexual identities. This point was made in Chapter 4 where it was revealed that participants did not understand their sexual selves in any simple way as traditionally masculine or feminine. In fact, their narratives often drew on dominant *and* resistant discourses of (hetero)sexuality indicating a complex accommodation and resistance of conventional ideas about male and female sexuality. It was also apparent that constitutions of sexual subjectivity were influenced by social context, so that what a young man revealed about his sexual self in front of other male peers may alter significantly from the sexual self his girlfriend sees. The implications of this for sexuality education are that any health promotion messages will need to take account of this subjective complexity and recognise its fluidity. Perpetually evolving meanings of male and female sexuality render ideas about young women as sexually passive and young men as sexually predatory 'old fashioned' for many in this study. Even if at the level of material reality this mode of gender sexual relations is still largely operative, at a perceptual level young people can see these identities as outdated. Targeting this perceptual sense of self even if this doesn't conform neatly to young people's lived reality, will be vital if sexuality education is to reach young people with its messages. Similarly, sexuality education may construct its messages in ways that speak to young people in recognition of the pressures they are under to present particular sexual identities. So for instance, building into safer sex teachings an acknowledgement that carrying condoms may make young women appear too sexually knowing but this may render them some control over whether a condom was produced.

The narratives of embodied sexual pleasure that appear throughout this book are testimony to the centrality of corporeality in young people's experience of sexuality. Leaving out the sensual body in sexuality education and only portraying a de-eroticised and medical physiology denies young people information about an essential component of sexuality. Including information in sexuality education about the potential pleasures of embodied sexual experience should be young people's right. Without this information about what feels pleasurable and what doesn't, young people and especially young women, have minimal knowledge upon which to base their decisions about engaging in sexual activity. The development of a spectrum of sexual embodiment in Chapter 5 also suggests these embodied states may have an important role to play in positive sexual experiences where young people exhibit respect for themselves and their partner. Given this, there may be some additional value in sexuality education promoting sexual embodiment in which young people are taught to value the sensual experience of their bodies rather than to ignore it. To do this however requires that they have access to information which acknowledges their bodies as sensuous and potentially pleasure-feeling.

Directions for future research

As this book has focused on building an argument for the inclusion of a discourse of erotics in sexuality education, working out the specifics of how this may operate within schools has been beyond its scope. Although I have discussed ideas about how this discourse might be configured, such as the need for it not to be heteronormative I have not delved into the pedagogical details of what might be taught and how. These decisions have to be made in accordance with the needs of particular students and the values, identities and understandings of any local context. On the basis of this, the content of the curriculum and the mode in which it is delivered will vary from school to school. The presence of a discourse of erotics in sexuality education does not mean that students will necessarily take up these meanings. Some students, especially those from religious and cultural backgrounds that oppose the kinds of meanings it advocates, may outrightly resist such a discourse. As has been emphasised throughout this book, young people do not passively absorb the things they are told but actively make meaning out of knowledge they receive. Including a discourse of erotics in sexuality education is about opening up possibilities for young people to experience themselves as sexual subjects in positive and self-determining ways. It should not be about replacing the limits of one discourse with another. Working out how a discourse of erotics may manifest in a particular school environment is the next step in any practical application of the research findings.

There are many questions left unanswered in relation to why some young people draw on resistant discourses of (hetero)sexuality when describing their sexual subjectivities and others don't. While the research suggests this is governed to some extent by access to resistant discourses, it is still not clear what encourages some and not all who have exposure to them, to think of themselves in these ways. Similarly, some young people are better able than others to 'work' these discourses for their own purposes. Some young men's identity-work in focus groups for example, enabled them to maintain an appropriate image of masculinity while participating in the seemingly unmasculine task of talking seriously about sexuality. This ability suggests the presence of a 'discursive literacy' that indicates skill at working discourses for one's own purposes. Further exploration of the idea of discursive literacy and how it might occur may shed additional light on the concept of agency.

While the research indicates that participant's experience of embodied sexuality is gendered what mediates this and its possible range of effects for sexual experience are still uncertain. Part of this enigma is tied to continuing debate about how subjectivity is conceptualised in terms of a nature/culture composition. Dominant discourses of (hetero)sexuality have a bearing on how we read experiences of the flesh but the nature of the effect of bodies on meaning is less well articulated. What are the possibilities of changing the documented lack of pleasure that many young women experience in sexual relationships? Would a change at the discursive level be enough to evoke a different material (i.e. enfleshed) experience for young women or does corporeality play an as yet unimagined part here? The spectrum of embodiment developed in Chapter 5 characterised the embodied experiences of young people in this research. However, I would be surprised if there weren't more sexually embodied states that could be inserted along this continuum with implications for gendered sexual experience and safer sex practice. What variables cause young people to vacillate along its nexus is an important line of investigation. Only further research into young people's sexually embodied experiences can draw out these nuances.

An underlying theme of this book has been the operation of power and how this is exercised at the intersections of subjectivity and discourse as well as through the micro-politics of young people's relationships. The study supports Foucault's assertion that power works productively and in more complex configurations than a monolithic conceptualisation of power allows. Young people's sense of being in control in situations and resisting the positions constituted within dominant discourses of heterosexuality can be understood as more than false consciousness. This is epitomised by the concept of mediated power employed in Chapter 6 to describe young women's feeling of agency in relationships. To take this theoretical discussion further would involve exploring how what are largely perceptual constructions of this agency are actually worked out in practice. The current study did not observe sexual decision making in the context of young people's everyday lives and so the relationship between perceived agency and actual agency could not be assumed. As previously explained, this

does not invalidate the findings as participants' discursively constituted narratives have real effects for the possibilities of their actions.
However one way of determining the 'revolutionary' potential of this
discursively conceived agency might be a study of its manifestation in
everyday practice.

It has been my desire that this study opens discursive space for alternative ways of thinking about young people's (hetero)sexuality and the
possibilities of sexuality education. It is with this sense of possibility
that I hope other researchers and those who work with young people
around issues of sexuality will continue future projects in this area.

Appendix   Couple Profiles

Couple

1. Amy and Peter	Amy is 17 and Peter is 18. They had been going out for over three years. They met at school where Peter asked Amy out. Both were virgins before they slept together. There was some discrepancy over when they first had sex, but this seemed to be 6 – 12 months into their relationship. Their relationship appeared fairly stable, although Amy expressed interest in meeting other potential partners. Both were still at school when I first met them.
2. Becky and Ashby	Both were 17 years old and still at school. They had been going out for three years and six months. They met at a school social and Ashby contacted Becky to ask her out the next day. It was eight months before they slept together and this was the first time either had experienced sexual intercourse. The first time they had sex Becky fell pregnant and she had an abortion. They had broken up several times during the course of the relationship and both times Ashby said he had slept with someone else at this time.
3. Nina and Neil	Were both unemployed and currently undertaking job training. Nina in secretarial work, and Neil in car mechanics. They had been going out for two years and initially met when Neil shouted something out of a car at Nina on her way to the beach. They had sex one year after meeting and both were virgins. Their relationship was as Neil puts it 'not calm' and they argued even in my presence. They had broken up once in this period.
4. Emma and Tim	Emma was 17 and still at school, while Tim was 18 and in his second year at University. They had been going out six months after meeting for the first time at a secondary school ball. Sexual intercourse occurred a few months into their relationship. While it was Tim's first sexual experience, Emma had several previous sexual partners. While Emma described the relationship as 'bliss', Tim appeared to feel less satisfied.

Couple – *continued*

5. Ngaire and George	Ngaire was 19, George was 21. Both were working. They had been going out 9.5 months after they met at a party of Ngaire's aunty. Although Ngaire had wanted to have sex on the night they met, they did not have intercourse until some weeks later. Both had experienced sexual intercourse before this encounter. They appeared to be extremely happy together.
6. Cam and Chris	This couple were both 19 and at University doing an Arts and Science degree consecutively. They met through mutual friends in the University Halls of Residence and had been going out three months. Both had experienced intercourse with other partners before sleeping together – this occurred on the first night they met. Their relationship appeared stable and they were contemplating moving in together next year.

Notes

Introduction

1 I use the word 'sex education' to refer to a historical form of programmes that were primarily concerned with biological and reproductive aspects of sexuality. Since 1999 these programmes have been known as sexuality education and are more holistic. See section in this chapter entitled 'Sexuality Education: The New Zealand Context'.

2 This project owes much to the Women Risk and AIDs and Men Risk and AIDs projects in Britain whose insights are the launching pad for this research.

3 Reports included the McMillian Inquiry (1937); Thomas Report (1945); Mazengarb Inquiry (1954); Currie Commission (1962); Ross Report (1973); Johnson Report (1977).

4 According to the *Health and Physical Education* curriculum Hauora is 'a Maori philosophy of health unique to New Zealand. It comprises Taha tinana (physical well-being), Taha hinengaro (mental and emotional well-being), Taha whanau (social well-being) and Taha wairua (spiritual well-being).

5 The *Health and Physical Education* curriculum states that 'through learning experiences that reflect the socio-ecological perspective, students can seek to remove barriers to healthy choices. They can help to create the conditions that promote their own well-being and that of other people and society as a whole' (Ministry of Education, 1999, p. 33).

6 Boards of Trustees have control over the executive decision making of individual schools in New Zealand. Their members are made up of the school's principal, parents, teachers and interested members of the public.

7 The concept of a discourse of erotics is explored in depth in Chapter 7.

8 This word means Maori and same-sex attracted.

2   Researching Sexuality: Methodological Complexities

1 Pakeha is a Maori term used to refer to non-Maori New Zealanders of European descent.

2 'The Ministry of Education uses a decile rating system for school funding purposes. Each decile contains approximately 10 per cent of schools. Schools in decile one have the highest proportion of students from low socio-economic backgrounds. Schools in decile 10 have the lowest proportion of these students' (www.minedu.govt.nz).

3 School Certificate examination is the first major examination at secondary school. Students sit it at around 15 years old (Form 5, Year 11). Sixth Form Certificate follows the year after.

4 Governing body of secondary schools in New Zealand composed of parents, teachers and members of the community interested in the schools' operation.

5 Pictures portraying dominant discourses of (hetero)sexuality consisted of a couple dancing romantically and another clinched in a passionate embrace. More alternative discourses of heterosexuality were communicated through an advert revealing a woman reaching through a man and dropping his heart on the floor and two pensioners sitting on a park bench kissing.
6 Statistical Package for the Social Sciences now known as SPSS Inc.
7 Palagi loosely translated means European.
8 Kaupapa Maori are Maori philosophies, values, principals and approaches. Definition from Glossary of Maori words Jones, Herda and Suaalii (2000).
9 In Walker's words 'Ka Whawhai Tonu Matou'.

3   Sperm Meets Egg?: Young People's Conceptualisations of Sexual Knowledge

1 Hui, Maori word for meeting, gathering, assembly.
2 Intermediate follows primary school and comes before secondary school in New Zealand. Students are aged 11 and 12 years at this stage.
3 Ninety-two per cent indicated they knew how a person can contract HIV/AIDS.
4 Eighty-eight per cent reported they knew how to put on a condom.
5 Thirteen per cent or 33 out of the 248 young people who answered this question mentioned this.
6 Thirty-eight per cent with 96 mentions from young people.
7 Twenty-two per cent (almost a quarter) with 54 mentions from young people.
8 In Wights' research with 14–16 year old males in a working class locality of Glasgow a survey question asking them to rank the three sources from which they got most of their information about sex revealed this order: Friends, Television, Parents.
9 Amy was the only exception here, and this may have been because at the time she started having sexual intercourse she had few friends to turn to. In her own words she explained 'I wasn't a very popular person, I didn't have many friends'.
10 Sig. 030.
11 Severe criticism of the inadequacy of school sex education from young men and women was also evident in the WRAP and MRAP projects. In this study young people dismissed the scientific model as: 'Latin names and mechanics', 'useless', 'rubbish', 'completely irrelevant, nothing to do with sex' (Holland et al., 1993, p. 7).
12 Sig. 000.
13 The actual percentages were 74 per cent of young women and 73 per cent of young men.
14 Forty per cent felt that knowledge did affect their relationships and ability to conduct them. These results are discussed in Chapter 6.
15 Sig. 037.

4   Sexual Subjects: Young People's Sexual Subjectivities

1 Sixty-seven per cent described themselves as 'sort of sexy' 10 per cent said they were 'very sexy'.

2 Fifty-six per cent said they had 'average desires' 30 per cent said they had 'very strong' desires.
3 Thirty-one per cent described themselves as 'fun loving', 29 per cent 'caring' and 27 per cent 'romantic'.
4 Four per cent described themselves as 'raunchy', 11 per cent 'kinky' 10 per cent 'lustful'.
5 There were several occasions when girlfriends appeared to rebuke their boyfriends for 'macho' comments. For instance, during a group discussion Gabby gave her boyfriend Theo a smack on his leg when he suggested that girls were 'catty' while 'blokes' were unconcerned with gossip.
6 *Married with Children* is an American sitcom which was screened on New Zealand television in the 1990s. It depicts a couple where the husband (Al Bundy) often rebuffs his wife's overtures for sexual activity.
7 See Gavey (1991); Patton and Mannison (1995); Tolman, Spencer, Rosen-Reynoso and Porche (2003).
8 These types of benefits of a relationship were mentioned in responses from focus group participants to the question 'Why get involved in a relationship'?
9 'Love' 30 per cent of mentions, 'respect' 29 per cent, 'commitment' 27 per cent.
10 Or at least known to each other.
11 Sexual pressure was not a topic on my question schedule, but rather raised by participants of their own accord.

5 'Like I'm floating somewhere ten feet in the air': Experiencing the Sexual Body

1 Carlos Spencer is a member of the All Black rugby team.

6 Desire, Pleasure, Power: Understanding Young People's Sexual Relationships

1 For international research in this area, see Holland et al. (1998) Tolman et al. (2003); Stewart (1999); Morris and Fuller (1999); Breakwell and Fife Schaw (1992).
2 See Braun et al. (2003) 18–50 years, Gavey and McPhillips (1999) 22–43 years, Gavey et al. (2001) 22–43 years, Potts (2002) 23–50 years.
3 More young men reported they had been in a relationship (sig. 012). More young women described themselves as currently in one (sig. 015).
4 A T-test conducted on these responses showed young women as 5.8021 and young men 5.8374 with no significant gender difference.
5 This was a highly significant gender difference (sig. 000).
6 There was a significant difference of (sig. 000) for young men dating partners younger than themselves and of the same age (sig. 000).
7 Young women were more likely to have been in their current relationship for 1–2 years (sig. 004) while young men only 1–3 months (sig. 049).
8 For example, an Australian self administered questionnaire for 18 and 19 year olds, reported that young women were less sexually experienced with regard to casual sexual partners than young men (Rodden et al., 1996).

9   Again in an attempt to disrupt the conflation of sexual activity with sexual intercourse, subjects were provided with the explanation that 'sexually active means engaging in petting and/or sexual intercourse with a partner'.
10   The significant difference was (sig. 007).
11   For exceptions see Tolman (2002) and Ussher and Mooney-Somers (2000).
12   Neil had been critically injured in a fight outside a pub one night. The experience had encouraged him to reflect on his priorities in life and as a consequence he gave up drinking and mixing with what he described as 'the wrong crowd'.

7   Constituting a Discourse of Erotics in Sexuality Education

1   By contrast, 74.5 per cent of young women had never consulted porno-graphic magazines as a source of information. Of those who had, only 3 per cent reported them very useful. It is possible this is because a discourse of erotics can be accessed from other sources such as women's magazines.

References

www.minedu.govt.nz

Aggleton, P. J., Homans, H., and Warwick, I. (1988). Young people's health beliefs and AIDS. In P. Aggleton and H. Homans (eds), *Social aspects of AIDS*. Lewes: The Falmer Press.

Aggleton, P., Ball, A., and Mane, P. (2000). Editorial: Young people, sexuality and relationships. *Sexual and Relationship Therapy*, 15(3), 213–20.

Aggleton, P. and Campbell, C. (2000). Working with young people – towards an agenda for sexual health. *Sexual and Relationship Therapy*, 15(3), 283–96.

Aggleton, P. and Homans, H. (eds) (1988). *Social aspects of AIDS*. Lewes: The Falmer Press.

Alldred, P., David, M., and Smith, P. (2003). Teachers' views of teaching sex education: Pedagogy and models of delivery. *Journal of Educational Enquiry*, 4(1), 80–96.

Allen, L. (2003). Power talk: Young people negotiating (hetero)sex. *Women's Studies International Forum*, 26(3), 235–44.

Allen, L. (in press). Concrete and classrooms: How schools shape educational research. *British Journal of Sociology of Education*.

Allen, L. (in press A). Trying not to think 'straight': Conducting focus groups with gay and lesbian youth. *International Journal of Qualitative Research*.

Barbour, R., Kitzinger, S., and Kitzinger, J. (eds) (1999). *Developing focus group research: Politics, theory and practice*. London: Sage.

Barker, G. (2000). Gender equitable boys in a gender inequitable world: Reflections from qualitative research and programme development in Rio de Janeiro. *Sexual and Relationship Therapy*, 15(3), 263–82.

Berne, L. and Huberman, B. (1999). *European approaches to adolescent sexual behaviour and responsibility – Executive summary and call to action*. Washington DC: Advocates for Youth.

Bhopal, K. (1997). *Gender, 'race' and patriarchy: A Study of South Asian Women*. Aldershot: Ashgate.

Blake, S. and Katrak, K. (2002). *Faith, Values and Sex and Relationships Education*. London: National Children's Bureau.

Blanc, A. (2001). The effect of power in sexual relationships on sexual and reproductive health: an examination of the evidence. *Studies in Family Planning*, 32(3), 189–213.

Bloor, M., Monaghan, L., Dobash, R., and Dobash, R. (1998). The body as a chemistry experiment: Steroid use among South Wales bodybuilders. In S. Nettleton and J. Watson (eds), *The body in everyday life*. London: Routledge.

Bollerud, K., Christopherson, S., and Frank, S. (1990). Girls' sexual choices: Looking for what is right. In C. Gilligan, N. Lyons, and T. Hamner (eds), *Making connections: The relational worlds of adolescent girls at the Emma Willard School*. Cambridge Mass: Harvard University Press.

Bowles, G. and Klein, R. (eds) (1983). *Theories of women's studies*. Boston: Routledge and Kegan Paul.

Brander, P. (1991). *Adolescent sexual practices: A study of sexual experiences and service needs among a group of New Zealand adolescents.* Wellington: Department of Health.

Braun, V., Gavey, N., and McPhillips, K. (2003). The 'fair deal'? Unpacking accounts of reciprocity in heterosex. *Sexualities, 6*(2), 237–61.

Breakwell, M. and Fife-Schaw, C. (1992). Sexual activities and preferences in a United Kingdom sample of 16 to 20-year-olds. *Archives of Sexual Behaviour, 21*(3), 271–93.

Britzman, D. (1998). *Lost subjects, contested objects: Towards a psychoanalytic inquiry of learning.* New York: State University of New York Press.

Brook, B. (1999). *Feminist perspectives on the body.* Harlow: Pearson Education Limited.

Butler, J. (1990). *Gender trouble: Feminism and the subversion of identity.* New York: Routledge.

Butler, J. (1993). *Bodies that matter: On the discursive limits of 'sex'.* London: Routledge.

Carroll, J., Volk, K., and Hyde, J. (1985). Differences between males and females in motives for engaging in sexual intercourse. *Archives of Sexual Behaviour, 14*(2), 131–39.

Coggan, C., Disley, B., Patternson, P., and Norton, R. (1997). Risk-taking behaviours in a sample of New Zealand adolescents. *Australian and New Zealand Journal of Public Health, 21*(5), 455–61.

Coleman, J. (1980). *The nature of adolescence.* London: Methuen.

Conley, L. (2003). We're here, we're queer: Get used to it. *SIECUS Report, 31*(4), 4–5.

Connell, R. (1987). *Gender and power.* Cambridge: Polity Press.

Connell, R. (1989). Cool guys, swots and wimps: The interplay of masculinity and education. *Oxford Review of Education, 15*(3), 291–303.

Connell, R. (1995). *Masculinities.* Cambridge: Polity Press.

Coward, R. (ed.) (1987). *Sexual violence and sexuality, feminist review: Sexuality a reader.* London: Virago.

Cram, F. (1997). Developing partnerships in research: Pakeha researchers and Maori research. *Sites, 35,* 44–63.

Csordas, T. (1990). Embodiment as a paradigm for anthropology. *Ethos, 18*(1), 5–47.

Csordas, T. (1994). *Embodiment and experience: The existential ground of culture and self.* Cambridge: Cambridge University Press.

Davies, B. (1989). *Frogs, snails and feminist tales: Preschool children and gender.* Sydney: Allen and Unwin.

Davies, B. (1993). *Shards of glass: Children reading and writing beyond gendered identities.* Sydney: Allen and Unwin.

Davies, B. (1997). The subject of post-structuralism: A reply to Alison Jones. *Gender and Education, 9*(3), 271–83.

Davies, D. (2000). Sex and relationship facilitation project for people with disabilities. *Sexuality and Disability, 18*(3), 187–94.

Davis, P. (ed.) (1996). *Intimate details and vital statistics: AIDS, sexuality and the social order in New Zealand.* Auckland: Auckland University Press.

DfE (1994). Education Act 1993: Sex education in schools (Circular 5/94). DfE.

Diamond, L. and Savin-Williams, R. (2000). Explaining diversity in the development of same-sex sexuality among young women. *Journal of Social Issues*, 56(2), 297–333.

Dickson, N. (1996). Sexual Behaviour. In P. Silva and W. Stanton (eds), *From Child to Adult: the Dunedin multidisciplinary health and development study*. Auckland: Oxford University Press.

Dickson, N., Paul, C., and Herbison, P. (1993). Adolescents, sexual behaviour and implications for an epidemic of HIV/AIDS among them. *Genitourinary Medicine*, 69, 133–40.

Dickson, N., Paul, C., Herbison, P., and Silva, P. (1998). First sexual intercourse: age, coercion, and later regrets reported by a birth cohort. *British Medical Journal*, 316, 29–33.

Duncombe, J. and Marsden, D. (1993). Love and intimacy: The gender division of emotion and emotion work. *Sociology*, 27(2), 221–41.

Dunne, M., Martin, N., Bailey, M., Heath, A., Bucholz, K., Madden, P., and Statham, D. (1997). Participation bias in a sexuality survey: Psychological and behavioural characteristics of responders and non-responders. *International Journal of Epidemiology*, 26(4), 844–54.

Dworkin, A. (1981). *Pornography: Men possessing women*. London: Women's Press.

Dworkin, A. (1987). *Intercourse*. New York: Free Press.

Eadie, J. (ed.) (2004). *Sexuality: The essential glossary*. London: Arnold.

Earle, S. (2003). Bumps and boobs: Fatness and women's experiences of pregnancy. *Women's Studies International Forum*, 26(3), 245–52.

Edley, N. and Wetherell, M. (1997). Jockeying for position: The construction of masculine identities. *Discourse and Society*, 8(2), 203–17.

Elliott, K. (1997). *Adolescent's perceptions of school based sexuality programmes*. Unpublished Masters Thesis: University of Auckland.

Elliott, K. (2003). The hostile vagina: Reading vaginal discourse in a school health text. *Sex Education*, 3(2), 133–44.

Epstein, D. and Johnson, R. (eds) (1998). *Schooling sexualities*. Buckingham: Open University Press.

Epstein, D. (1994). *Challenging lesbian and gay inequalities in education*. Milton Keynes: Open University Press.

Epstein, D., O'Flynn, S., and Telford, D. (2003). *Silenced Sexualities in Schools and Universities*. London: Trentham Books Limited.

Featherstone, M. (1991). The body in consumer culture. In M. Featherstone, M. Hepworth and B. Turner (eds), *The body: Social processes and cultural theory*. London: Sage.

Fine, M. (1988). Sexuality, schooling and adolescent females: the missing discourse of desire. *Harvard Educational Review*, 58(1), 29–51.

Fine, M. (1994). Working the Hyphens: Reinventing self and other in qualitative research. In N. Denzin and Y. Lincoln (eds), *Handbook of qualitative research*. California: Sage.

Fine, M. and Weis, L (1996). Writing the 'wrongs' of fieldwork: Confronting our own research/writing dilemmas in urban ethnographies. *Qualitative Inquiry*, 2(3), 251–74.

Flax, J. (1992). The end of innocence. In J. Butler and Scott, J. (eds), *Feminists theorize the political*. London: Routledge.

Forrest, S. (2000). Big and tough: Boys learning about sexuality and manhood. *Sexual and Relationship Therapy, 15*(3), 247–261.

Foucault, M. (1976). *The history of sexuality, Volume 1.* trans. R. Hurley, Harmondsworth: Penguin.

Foucault, M. (1977). *Discipline and punish: The birth of the prison.* Trans. A Sheridan. London: Allen Lane.

Foucault, M. (1980). The history of sexuality: An interview (trans. Geoff Benington). *Oxford Literary Review, 4*(2), page numbers

Foucault, M. (1980a). *Power/knowledge: Selected interviews and other writings 1972–1977.* London: Harvester Wheatsheaf.

Foucault, M. (1983). Afterword: The subject and power. In D. Hubert and P. Rabinow (eds), *Michel Foucault: Beyond structuralism and hermeneutics.* Chicago: The University of Chicago Press.

Francis, B. and Skelton, C. (eds) (2001). *Investigating gender: Contemporary perspectives in education.* Buckingham: Open University Press.

Frosh, S., Phoenix, A., and Pattman, R. (2002). *Young masculinities.* Basingstoke: Palgrave.

Fuss, D. (1989). *Essentially speaking: Feminism, nature and difference.* New York: Routledge.

Gagnon, J. and Simon, W. (1973). *Sexual conduct: The social sources of human sexuality.* Chicago: Aldine.

Gatens, M. (1996). *Imaginary bodies: Ethics, power and corporeality.* London: Routledge.

Gavey, N. (1992). Technologies and effects of heterosexual coercion. *Feminism and Psychology, 2*(3), 325–51.

Gavey, N. and McPhillips, K. (1999). Subject to romance: Heterosexual passivity as an obstacle to women initiating condom use. *Psychology of Women Quarterly, 23*(2), 325–51.

Gavey, N., McPhillips, K., and Doherty, M. (2001). If it's not on, it's not on – or is it?: Discursive constraints on women's condom use. *Gender and Society, 15*(6), 917–34.

Gray, J. (1995). *Mars and Venus in the bedroom: A guide to lasting romance and passion.* New York: HarperCollins.

Grogan, S. and Richards, H. (2002). Body image: Focus groups with boys and men. *Men and Masculinities, 4*(3), 219–32.

Grosz, E. (1987). Notes towards a corporeal feminism. *Australian Feminist Studies: Special Issue, Feminism and the Body, 5,* 1–16.

Grosz, E. (1994). *Volatile bodies: Towards a corporeal feminism.* Bloomington: Indiana University Press.

Guggino, J. and Ponzetti, J. (1997). Gender differences in affective reactions to first coitus. *Journal of Adolescence, 20*(2), 189–200.

Gurevich, M., Bishop, S., Bower, J., Malka, M., and Nyhof-Young, J. (2004). (Dis)embodying gender and sexuality in testicular cancer. *Social Science and Medicine, 58,* 1597–1601.

Harding, S. (1993). Rethinking standpoint epistemologies: What is strong objectivity? In L. Alcoff and E. Potter (eds), *Feminist epistemologies.* London: Routledge.

Harris, A., Aapola, S., and Gonick, M. (2000). Doing it differently: Young women managing heterosexuality in Australia, Finland, and Canada. *Journal of Youth Studies, 3*(4), 373–88.

Hawkes, G. (1996). *A sociology of sex and sexuality*. Buckingham: Open University Press.

Haywood, C. (1996). Out of the curriculum: sex talking, talking sex. *Curriculum Studies*, 4(2), 229–49.

Hearn, J. and Morgan, D. (1995). Contested discourses on men and masculinities. In J. Holland, M. Blair, and S. Sheldon (eds), *Identity and diversity: Gender and the experience of education*. London: Open University.

Hekman, S. (1995). Subjects and agents: The question for feminism. In J. Kegan Gardiner (ed.), *Provoking agents: Gender and agency in theory and practice*. Chicago: University of Illinois Press.

Henriques, J., Hollway, W., Urwin, C., Venn, C., and Walkerdine, V. (1984). *Changing the subject: Psychology, social regulation and subjectivity*. London: Methuen.

Herek, G. (1987). On heterosexual masculinity: some psychological consequences of the social construction of gender and sexuality. In M. Kimmel (ed.), *Changing men: New directions in research on men and masculinity*. Newbury Park, California: Sage.

Hertz, R. (ed.) (1997). *Reflexivity and voice*. Thousand Oaks, CA: Sage.

Hey, V. (1997). *The company she keeps, an ethnography of girls' friendships*. Buckingham: Open University Press.

Hillier, L., Dempsey, D., Harrison, L., Beale, L., Matthews, L., and Rosenthal, D. (1998). *Writing themselves in: A national report on the sexuality, health and well-being of same-sex attracted young people*. Monograph series 7. La Trobe University.

Hillier, L., Harrison, L., and Bowditch, K. (1999). Neverending love and Blowing your load: The meanings of sex to rural youth. *Sexualities*, 2(1), 69–88.

Hingson, R. and Strunin, L. (1992). Monitoring adolescent's responses to the AIDS epidemic: Changes in knowledge, attitudes, beliefs and behaviours. In R. DiClemente (ed.), *Adolescents and AIDS: A generation in jeopardy*. Newbury Park, Calif: Sage.

Holibar, F. (1992). *A qualitative investigation into teenage relationships*. Alcohol and Public Health Research Unit, School of Medicine. Auckland.

Holland, J., Ramazanoglu, C., Scott, S., Sharpe, S., and Thomson, R. (1991). *Pressure, resistance, empowerment: Young women and the negotiation of safer sex*. WRAP Paper 6. London: Tufnell Press.

Holland, J., Ramazanoglu, C., Scott, S., Sharpe, S., and Thomson, R. (1991a). Between embarrassment and trust: Young women and the diversity of condom use. In P. Aggleton, G. Hart, and P. Davies (eds), *AIDS: Responses, interventions and care*. London: Falmer Press.

Holland, J., Ramazanoglu, C., and Sharpe, S. (1993). *Wimp or gladiator: Contradictions in acquiring masculine sexuality*. Women, risk and AIDS project, men risk and AIDS project. London: Tufnell Press.

Holland, J., Ramazanoglu, C., Sharpe, S., and Thomson, R. (1994). Achieving masculine sexuality: Young men's strategies for managing vulnerability. In L. Doyal, J. Naidoo, and T. Wilton (eds), *AIDS setting a feminist agenda*. London: Taylor and Francis.

Holland, J., Ramazanoglu, C., Sharpe, S., and Thomson, R. (1994a). Power and desire: The embodiment of female sexuality. *Feminist Review*, 46, 21–38.

Holland, J., Ramazanoglu, C., Sharpe, S., and Thomson, R. (1996). Reputations: Journeying into gendered power relations. In J. Weeks and J. Holland (eds), *Sexual cultures: Communities, values, and intimacy*. London: Macmillan – now Palgrave.

Holland, J., Ramazanoglu, C., Sharpe, S., and Thomson, R. (1998). *The male in the head: Young people, heterosexuality and power*. London: Tufnell Press.

Holliday, R. and Hassard, J. (2001). *Contested bodies*. London: Routledge.

Hollway, W. (1984). Gender difference and the production of subjectivity. In J. Henriques, W. Hollway, C. Urwin, C. Venn, and V. Walkerdine (eds), *Changing the subject: Psychology, social regulation and subjectivity*. London: Methuen.

Hollway, W. (1989). *Subjectivity and method in psychology: Gender, meaning and science*. London: Sage.

Ineson, S. (1996). Importance of sex education. *Evening Post*. 7 September.

Ingraham, C. (2002). The heterosexual imaginary. In S. Jackson and S. Scott (eds), *Gender: A sociological reader*. London: Routledge: London.

Innocenti (2001). *A league table of teenage births in rich countries: Innocenti report card issue 3*. Florence: UNICEF.

Jackson, M. (1984). Sex research and the construction of sexuality: A tool for male supremacy? *Women's Studies International Forum, 7*, 43–51.

Jackson, S. (2001). Happily never after: Young women's stories of abuse in heterosexual love relationships. *Feminism and Psychology, 11*(3), 305–21.

Jackson, S. and Scott, S. (eds) (1996). *Feminism and sexuality: A reader*. Edinburgh University Press.

Jeffreys, S. (1996). Heterosexuality and the desire for gender. In D. Richardson (ed.), *Theorising heterosexuality*. Buckingham: Open University Press.

Jones, A. (1992). Writing feminist educational research: Am 'I' in the text?. In S. Middleton and A. Jones (eds), *Women and education in Aotearoa 2*. Wellington: Bridget Williams Books.

Jones, A. (1993). Becoming a 'girl': Post-structuralist suggestions for educational research. *Gender and Education, 5*(2), 157–66.

Jones, A. (1997). Teaching post-structuralist feminist theory in education: Student resistances. *Gender and Education, 9*(3), 261–69.

Jones, A. (1999). The limits of cross-cultural dialogue: Pedagogy, desire and absolution in the classroom. *Educational Theory, 49*(3), 299–316.

Jones, A., Herda, P., and Suaalii, T. (eds) (2000). *Bitter sweet: Indigenous women in the Pacific*. Dunedin: University of Otago Press.

Kehily, M. (2001). Bodies in school, young men, embodiment and heterosexual masculinities. *Men and Masculinities, 4*(2), 173–85.

Kehily, M. (2002). *Sexuality, gender and schooling: Shifting agendas in social learning*. London: Routledge Falmer.

Kehily, M. and Nayak, A. (1997). Lads and laughter: Humour and the production of heterosexual hierarchies. *Gender and Education, 9*(1), 69–87.

Kelly, L., Burton, S., and Regan, L. (1994). Researching women's lives or studying women's oppression? Reflections on what constitutes feminist research. In M. Maynard and J. Purvis (eds), *Researching women's lives from a feminist perspective*. London: Taylor and Francis.

Kenway, J. and Willis, S. (1997). *Answering back: Girls, boys and feminism in schools*. Sydney: Allen and Unwin.

Kitzinger, C. and Wilkinson, S. (1994). Virgins and queers: Rehabilitating heterosexuality? *Gender and Society, 8*(3), 444–63.

Lather, P. (1991). *Getting smart: Feminist research and pedagogy with/in the post modern*. London: Routledge.

Lecourt, D. (1975). *Marxism and epistemology*. London: National Labour Board.

Leder, D. (1990). *The absent body*. Chicago, IL: University of Chicago Press.

Lee, R. (1993). *Doing research on sensitive topics*. London: Sage.

Lees, S. (1993). *Sugar and spice: Sexuality and adolescent girls*. London: Penguin.

Letherby, G. (2003). *Feminist research in theory and practice*. Buckingham: Open University Press.

Longhurst, R. (2001). Breaking corporeal boundaries: Pregnant bodies in public places. In R. Holliday and J. Hassard (eds), *Contested bodies*. London: Routledge.

Lupton, D. (1998). Going with the flow: Some central discourses in conceptualising and articulating embodiment of emotional states. In S. Nettleton and J. Watson (eds), *The body in everyday life*. London: Routledge.

Mac an Ghaill, M. (1994). *The making of men: Masculinities, sexualities and schooling*. Buckingham. Open University.

Mac an Ghaill, M. (ed.) (1996). *Understanding masculinities*. Buckingham: Open University Press.

Mac an Ghaill, M. (1996a). Deconstructing heterosexualities within school arenas. *Curriculum Studies, 4*(2), 191–209.

MacKinnon, C. (1996). Feminism, marxism, method and the State: An agenda for theory. In S. Jackson and S. Scott (eds), *Feminism and sexuality a reader*. Edinburgh: Edinburgh University Press.

MacKinnon, K. (1989). *Towards a feminist theory of the state*. Cambridge MA: Harvard University Press.

Magdala, P. (2002). Adolescent boys and the muscular male body ideal. *Journal of Adolescent Health, 30*(4), 233–42.

Malson, H. (1998). *The thin woman: Feminism, post-structuralism and the social psychology of anorexia nervosa*. London: Routledge.

McCabe, M. (1999). Sexual knowledge, experience and feelings among people with disability. *Sexuality and Disability, 17*(2), 157–70.

McKay, A. (1997). Accommodating Ideological Pluralism in Sexuality Education. *Journal of Moral Education, 26*(3), 285–300.

McPhillips, K., Braun, V., and Gavey, N. (2001). Defining (hetero)sex: How imperative is the 'coital imperative'?. *Women's Studies International Forum, 24*(2), 229–40.

McRobbie, A. (1991). *Feminism and youth culture: From 'Jackie' to 'Just Seventeen'*. London: Macmillan – now Palgrave.

McRobbie, A. (1996). More!: New sexualities in girls' and women's magazines. In J. Curran, D. Morley, and V. Walkerdine (eds), *Cultural studies and communications*. London: Arnold.

Measor, L., Tiffin, C., and Miller, K. (2000). *Young People's Views on Sex Education: Education, Attitudes and Behaviour*. London: Routledge Falmer.

Merleau-Ponty, M. (1962). *Phenomenology of perception,* Translated from French by Colin Smith. London: Routledge and Kegan Paul.

Middleton, S. (1995). Doing feminist educational theory – a post-modernist perspective. *Gender and Education, 7*(1), 87–100.

Middleton, S. (1998). *Disciplining sexuality – Foucault, life-histories and education*. New York: Teacher's College Press.

Milligan, M. and Neufeldt, A. (2001). *Sexuality and Disability, 9*(2), 91–109.

Milton, J., Berne, L., Peppard, J., Patton, W., Hunt, L., and Wright, S. (2001). Teaching sexuality education in high schools: what qualities do Australian teachers value? *Sex Education, 1*(2), 175–86.

Ministry of Education (1999). *Health and physical education in the New Zealand curriculum*. Wellington: Learning Media Limited.

Ministry of Health (1997). *Statement on sexual and reproductive health strategy*. Wellington: New Zealand Ministry of Health.

Ministry of Health (2001). *Sexual and reproductive health strategy: Phase One*. Wellington: New Zealand Ministry of Health.

Molitor, F., Facer, M., and Ruiz, J. (1999). Safer sex communication and unsafe sexual behaviour among young men who have sex with men in California. *Archives of Sexual Behaviour, 28*(4), 335–48.

Moore, S. and Rosenthal, D. (1998). Contemporary youths' negotiations of romance, love, sex and sexual disease. In V. De Munck (ed.), *Romantic love and sexual behaviour: Perspectives from the social sciences*. Westport, USA: Praeger Publishers.

Morris, K. and Fuller, M. (1999). Heterosexual relationships of young women in a rural environment. *British Journal of Sociology of Education, 20*(4), 531–43.

Munro, J., Ballard, V. (2004). Oh, what would you do Mrs Brown? Some Experiences in Teaching about Sexuality. *New Zealand Journal of Educational Studies, 39*(1), 71–90.

Nash, R. (2001). Sex and the school: The lessons of experience. *Social Work Review* Winter, 27–32.

Nayak, A. and Kehily, M. (1996). Playing it straight: masculinities homophobias and schooling. *Journal of Gender Studies, 5*(2), 211–30.

Nettleton, S. and Watson, J. (1998). The body in everyday life: An introduction. In S. Nettleton and J. Watson (eds), *The body in everyday life*. London: Routledge.

O'Donnell, M. and Sharpe, S. (2000). *Uncertain masculinities: Youth, ethnicity and class in contemporary Britain*. London: Routledge Falmer.

Ogden, J. (1996). *Health psychology: A textbook*. Buckingham: Open University Press.

Olesen, V. (2000). Feminisms and qualitative research into the millennium. In N. Denzin and Y. Lincoln (eds), *Handbook of qualitative research* 2nd Edition. Thousand Oaks: Sage.

Pan American Health Organisation and World Health Organisation (2000). *Promotion of sexual health recommendations for action: Proceedings of a regional consultation convened by the Pan American Health Organisation, World Health Organisation in collaboration with the World Association for Sexology*. Antigua, Guatemala.

Petelo, L. (1997). *Researching my other, my self*. Paper presented to the New Zealand Association for Research in Education Conference. Auckland.

Potts, A. (2002). *The science/fiction of sex: Feminist deconstruction and the vocabularies of heterosex*. London: Routledge.

Quinlivan, K. (1996). Claiming an identity they taught me to despise: lesbian students respond to the regulation of same sex desire. *Women's Studies Journal, 12*(2), 99–113.

Quinlivan, K. and Town, S. (1999). Queer pedagogy, educational practice and lesbian and gay youth. *Qualitative Studies in Education, 12*(5), 509–24.

Ramazanoglu, C. and Holland, J. (2002). *Feminist methodology: Challenges and choices*. London: Sage.

Redman, P. (1996). Empowering men to disempower themselves: Heterosexual masculinities, HIV and the contradictions of anti-oppressive education. In M. Mac an Ghaill (ed.), *Understanding masculinities*. Buckingham: Open University Press.

Redman, P. (2001). The discipline of love: Negotiation and regulation in boys' performance of romance-based heterosexual masculinity. *Men and Masculinities, 4*(2), 186–200.

Reiss, M. (1997). Editorial: The value(s) of sex education. *Journal of Moral Education, 26*(3), 253–55.

Rhoads, R. (1997). Crossing sexual orientation borders: collaborative strategies for dealing with issues of positionality and representation. *Qualitative Studies in Education, 10*(1), 7–23.

Richardson, D. (1996). Heterosexuality and social theory. In D. Richardson (ed.), *Theorising heterosexuality*. Buckingham: Open University Press.

Roberts, C., Kippax, S., Waldby, C. and Crawford, J. (1995). Faking it: The story of 'Ohh!'. *Women's Studies International Forum, 18*(5/6), 523–32.

Roberts, H. (ed.) (1981). *Doing feminist research*. London: Routledge and Kegan Paul.

Rodden, P., Crawford, J., Kippax, S., and French, J. (1996). Sexual practice and understandings of safe sex: Assessing change among 18–19 year old Australian tertiary students, 1988 to 1994. *Australian and New Zealand Journal of Public Health, 20*(6), 643–649.

Ryan, A. (1988). The 'moral right', sex education, and populist moralism. In S. Middleton (ed.), *Women and education in Aotearoa*. Wellington: Allen and Unwin.

Savin-Williams, R. (2001). A critique of research on sexual-minority youths. *Journal of Adolescence, 24*(1), 5–13.

Scott, S. and Morgan, D. (1993). *Body matters*. London: Falmer Press.

Sears, J. (ed.) (1992). *Sexuality and the curriculum*. New York: Teachers College Press.

Shakespeare, T., Gillespie-Sells, K., and Davies, D. (1996). *The sexual politics of disability: Untold desires*. London: Cassell.

Shilling, C. (ed.) (1993). *The body and social theory*. London: Sage.

Simon, J. (1994). Historical perspectives on education in New Zealand. In E. Coxon, K. Jenkins, J. Marshall, and L. Massey (eds), *The politics of learning and teaching in Aotearoa New Zealand*. Palmerston North: Dunmore Press.

Sinclair, J. (1995). *Collins English Dictionary*. Glasgow: HarperCollins.

Skeggs, B. (1997). *Formations of class and gender, becoming respectable*. London: Sage.

Smith, G. (1990). *Research issues related to Maori education*. Paper presented at the New Zealand Association for Research in Education Special Interests Conference. Education Department, University of Auckland.

Smith, L. (1992). Maori women: Discourses, projects, and Mana Wahine. In S. Middleton and A. Jones (eds), *Women and Education in Aotearoa 2*. Wellington: Bridget Williams.

Smyth, H. (2000). *Rocking the cradle: Contraception, sex, and politics in New Zealand*. Wellington: Steele Roberts Ltd.

Social Exclusion Unit (1999). *Report on teenage pregnancy*, HMSO Cm 4342.

SSRU (1994). Reviews of effectiveness: HIV prevention and sexual health interventions. Social Science Research Unit, Institute of Education, University of London, *SSRU Database Project No. 1*, September 1994 (revised edition).

Stanley, L. (1992). *Is there a lesbian epistemology?* Manchester: Feminist Praxis, Department of Sociology, University of Manchester.

Stanley, L. (ed.) (1990). *Feminist praxis: Research, theory and epistemology in feminist sociology.* London: Routledge.

Statistics New Zealand (2002). *2001 census of population and dwellings: Ethnic groups.* Wellington: Statistics New Zealand Te Tari Tatau.

Stenberg, S. (2002). Embodied classrooms, embodied knowledges: Re-thinking the mind/body split. *Composition Studies, 30*(2), 43–60.

Stewart, F. (1999). Femininities in flux? Young women, heterosexuality, and (safe) sex. *Sexualities, 2*(3), 275–90.

Tasker, G. (2004). Health Education: Contributing to a Just Society through Curriculum Change, In Anne-Marie O'Neill, John Clark and Roger Openshaw (eds), *Reshaping Culture, knowledge and Learning? Policy and Content in the New Zealand Curriculum Framework.* Volume One. Palmerston North: Dunmore Press.

Tepper, M. (2000). Sexuality and disability: The missing discourse of pleasure. *Sexuality and Disability, 18*(4), 284–90.

Thompson, S. (1990). Putting a big thing into a little hole: Teenage girl's accounts of sexual initiation. *Journal of Sex Research, 23*(3), 341–61.

Thomson, R. (1997). Diversity, Values and Social Change: renegotiating a consensus on sex education. *Journal of Moral Education, 26*(3), 257–71.

Thomson, R. (2000). Dream on: The logic of sexual practice. *Journal of Youth Studies, 3*(4), 407–27.

Thomson, R. and Scott, S. (1991). Learning about sex: Young women and the social construction of sexual identity. *Women, Risk and AIDS Project: Young women, sexuality and the limitation of AIDS*, Paper 4. London: Tufnell Press.

Thomson, R., McGrellis, S., Holland, J., Henderson, S., and Sharpe, S. (2001). From 'Peter Andre's six pack' to 'I do knees' – the body in young people's moral discourse. In K. Backett and L. Mckie (eds), *Constructing gendered bodies.* Basingstoke: British Sociological Association.

Thorogood, N. (2000). Sex education as disciplinary technique: Policy and practice in England and Wales. *Sexualities, 3*(4), 425–38.

Tolman, D. (1994). Daring to desire: Culture and the bodies of adolescent girls. In J. Irvine (ed.), *Sexual cultures the construction of adolescent identities.* Philadelphia: Temple University Press.

Tolman, D. (2002). *Dilemmas of desire: Teenage girls talk about sexuality.* Harvard: Harvard University Press.

Tolman, D. and Higgins, T. (1996). How being a good girl can be bad for girls. In N. Maglin and D. Perry (eds), *'Bad girls' 'Good girls': Women, sex, and power in the nineties.* New Jersey: Rutgers University Press.

Tolman, D., Spencer, R., Rosen-Reynoso, M., and Porche, M. (2003). Sowing the seeds of violence in heterosexual relationships: Early adolescents narrate compulsory heterosexuality. *Journal of Social Issues, 59*(1), 159–78.

Turner, B. (1992). *Regulating bodies: Essays in medical sociology.* London: Routledge.

Turner, B. (1996). *The body and society,* 2nd Edition. London: Sage.

Ussher, J. and Mooney-Somers, J. (2000). Negotiating desire and sexual subjectivity: Narratives of young lesbian avengers. *Sexualities, 3*(2), 183–200.

Vincent, K. and Ballard, K. (1997). Living on the margins: Lesbian experience in secondary schools. *New Zealand Journal of Educational Studies, 32*(2), 147–61.

Waldby, C., Kippax, S., and Crawford, J (1993). Cordon Sanitaire: 'Clean' and 'unclean' women in the AIDS discourse of young heterosexual men. In P. Aggleton, P. Davies and G. Hart (eds), *AIDS: Facing the second decade.* London: Falmer.

Walker, B. and Kushner, S. (1997). Understanding boys' sexual health education and its implications for attitude change. Final report of research funded by E.S.R.C. Centre for Applied Research in Education, University of East Anglia.

Walker, R. (1990). *Ka Whawhai Tonu Matou: Struggle without end.* Auckland: Penguin.

Walkerdine, V. (1984). The insertion of Piaget into early childhood education. In J. Henriques, W. Hollway, C. Urwin, C. Venn, and V. Walkerdine (eds), *Changing the subject: Psychology, social regulation, and subjectivity.* London: Methuen.

Warr, D. (2001). The importance of love and understanding: Speculation on romance in safe sex health promotion. *Women's Studies International Forum, 24*(2), 241–52.

Watson, J. (2000). *Male bodies, health, culture, and identity.* Buckingham: Open University Press.

Weedon, C. (1987). *Feminist practice and post-structuralist theory.* Oxford: Blackwell.

Weeks, J. (1989). AIDS: The intellectual agenda. In P. Aggleton, G. Hart, and P. Davies (eds), *AIDS: Social representations, social practices.* Lewes: Falmer Press.

Weeks, J. and Holland, J. (eds) (1996). *Sexual cultures: Communities, values, and intimacy.* New York: St. Martin's Press.

Weeks, J., Holland, J., and Waites, M. (eds) (2003). *Sexualities and society: A reader.* Cambridge: Polity Press.

Weitz, R. (ed.) (1998). *The politics of women's bodies: Sexuality, appearance, and behaviour.* New York: Oxford University Press.

Wellings, K. and Field, B. (1996). *Stopping AIDS: AIDS/HIV public education and the mass media in Europe.* London: Longman.

Wetherell, M. and Edley, N. (1998). Gender practices: Steps in the analysis of men and masculinities. In K. Henwood, C. Griffin, and A. Phoenix (eds), *Standpoints and differences: Essays in the practice of feminist psychology.* London: Sage.

Wight, D. (1994). Boys' thoughts and talk about sex in a working class locality of Glasgow. *Sociological Review, 42*(4), 703–37.

Williams, S. (1996). The vicissitudes of embodiment across the chronic illness trajectory. *Body and Society, 2*(2), 23–47.

Willig, C. (1999). Discourse Analysis and sex education. In C. Willig (ed.), *Applied discourse analysis: Social and psychological interventions.* Philadelphia: Open University Press.

Willig, C. (2001). *Introducing qualitative research in psychology: Adventures in theory and method.* Buckingham: Open University Press.

Willis, P. (1977). *Learning to Labour*. Farnborough: Saxon House.

Witz, A. (2000). Whose body matters? Feminist sociology and the corporeal turn in sociology and feminism. *Body and Society, 6*(2), 1–24.

Wood, J. (1984). Groping towards sexism: Boy's sex talk. In A. McRobbie and M. Nava (eds), *Gender and generation*. London: Macmillan – now Palgrave.

Copyright Acknowledgements

Versions of chapters from this book have been originally published elsewhere. I am grateful to the following for permission to reproduce this material:

Sage Publications for material from Allen, L. (2003) 'Girls Want Sex, Boys Want Love: Resisting Dominant Discourses of (Hetero)sexuality', *Sexualities*, 6(2): 215–36; Allen, L. (2005) 'Managing Masculinity: Young Men's Identity Work in Focus Groups', *Qualitative Research*, 5(1): 35–57.

Carfax Publishing, Taylor & Francis Ltd for material from Allen, L. (2001) 'Closing Sex Education's Knowledge/Practice Gap: the reconceptualisation of young people's sexual knowledge', *Sex Education*, 1(2): 109–22. http://www.tandf.co.uk/journals; for material from Allen, L. (2004) 'Beyond the Birds and the Bees: constituting a discourse of erotics in sexuality education', *Gender and Education*, 16(2): 151–67. http://www.tandf.co.uk/journals; for material from Allen, L. (2004) 'Getting Off and Going Out: Young people's Conceptualisations of (hetero)sexual relationships', *Culture, Health and Sexuality*, 6(6): 463–81. http://www.tandf.co.uk/journals

University of Otago Press for material from Allen, L. (2002) 'As far as sex goes I don't really think about my body': Young men's corporeal experiences of (hetero)sexual pleasure' in Worth, H., Allen, L., Paris, A. (eds) *The Life of Brian: Masculinities, Sexualities and Health in New Zealand* and for material from Allen, L. (2002) 'Naked Skin Together: Exploring young women's narratives of (hetero)sexual pleasure through a spectrum of Embodiment', *Women's Studies Journal* 18(11): 45–67.

Elsevier Science Ltd for material from Allen, L. (2003) 'Power Talk: Young People Negotiating (Hetero)sex', *Women's Studies International Forum*, 26(3): 235–44.

Index